Introductory Statistics

Introductory Statistics

Nancy Maxwell

Willford Press
www.willfordpress.com

Published by Willford Press,
118-35 Queens Blvd., Suite 400,
Forest Hills, NY 11375, USA

ISBN: 978-1-64728-351-3

Cataloging-in-Publication Data

Introductory statistics / Nancy Maxwell.
p. cm.
Includes bibliographical references and index.
ISBN 978-1-64728-351-3
1. Statistics. 2. Mathematical statistics. I. Maxwell, Nancy.
QA276 .I58 2022
519.5--dc23

For information on all Willford Press publications
visit our website at www.willfordpress.com

Willford Press

Table of Contents

Preface

This book aims to help a broader range of students by exploring a wide variety of significant topics related to this discipline. It will help students in achieving a higher level of understanding of the subject and excel in their respective fields. This book would not have been possible without the unwavered support of my senior professors who took out the time to provide me feedback and help me with the process. I would also like to thank my family for their patience and support.

The mathematical discipline that is concerned with the collection, analysis, organization, interpretation and presentation of data is referred to as statistics. Descriptive statistics and inferential statistics are the main statistical methods that are used in data analysis. Descriptive analysis uses indexes such as mean and standard deviation to summarize data from a sample. Distribution and dispersion are the two most important sets of properties of descriptive statistics. Inferential statistics uses data analysis to conclude the properties of the fundamental probability distribution. The topics included in this book on statistics are of utmost significance and bound to provide incredible insights to readers. It aims to shed light on some of the unexplored aspects of this field. Those in search of information to further their knowledge will be greatly assisted by this book.

A brief overview of the book contents is provided below:

Chapter - What is Statistics?

Statistics deals with the collection, organization, analysis and presentation of data through the use of quantified models and representations. The analyzed data uses two statistical methods - descriptive and inferential statistics. This is an introductory chapter which will briefly introduce about statistics.

Chapter - Branches of Statistics

The discipline of statistics can be categorized into various branches such as descriptive analysis, parametric and nonparametric statistics, exact statistics, etc. This chapter closely examines these branches of statistics to provide an extensive understanding of the subject.

Chapter - Statistical Measures

Statistical measures refer to the individual quantitative variable values for the statistical units in a specific group. Such measures include statistical mean, mode, median, range, skewness, quantile, quartile, variance, quartile deviation, pooled variance, standard deviation, etc. The topics elaborated in this chapter will help in gaining a better perspective of these statistical measures.

Chapter - Sampling Distributions

Sampling distribution refers to the probability distribution of data obtained from a large number of samples. Sampling distribution of mean, median, mode and standard deviation are studied within statistics. This chapter sheds light on the sampling distributions for an in-depth understanding of the subject.

Chapter - Statistical Inference

Statistical inference is the process that makes use of data analysis for deducing properties of a probability distribution. Algorithmic inference, fiducial inference and Bayesian inference fall under its domain. This chapter closely examines the varied aspects of statistical inference to provide an extensive understanding of the subject.

Chapter - Theorems in Statistics

Central limit theorem, Basu's theorem, Cochran's theorem, Fieller's theorem, Fisher- Tippett-Gnedenko theorem, Hajek-Le Cam convolution theorem, Neyman-Pearson lemma, etc. are some of the theorems that are used in statistics. This chapter discusses these theorems of statistics in detail.

Chapter - Applications

Statistics has applications in the fields of actuarial science, business analytics, forensics, finance, engineering, operations research, signal processing, psychology, machine learning, etc. This chapter has been carefully written to provide an easy understanding of the diverse applications of statistics.

Nancy Maxwell

1

What is Statistics?

Statistics deals with the collection, organization, analysis and presentation of data through the use of quantified models and representations. The analyzed data uses two statistical methods – descriptive and inferential statistics. This is an introductory chapter which will briefly introduce about statistics.

Statistics is a form of mathematical analysis that uses quantified models, representations and synopses for a given set of experimental data or real-life studies. Statistics studies methodologies to gather, review, analyze and draw conclusions from data. Some statistical measures include the following:

- Mean,
- Regression analysis,
- Skewness,
- Kurtosis,
- Variance,
- Analysis of variance.

Statistics is a term used to summarize a process that an analyst uses to characterize a data set. If the data set depends on a sample of a larger population, then the analyst can develop interpretations about the population primarily based on the statistical outcomes from the sample. Statistical analysis involves the process of gathering and evaluating data and then summarizing the data into a mathematical form.

Statistics is used in various disciplines such as psychology, business, physical and social sciences, humanities, government, and manufacturing. Statistical data is gathered using a sample procedure or other method. Two types of statistical methods are used in analyzing data: Descriptive statistics and inferential statistics. Descriptive statistics are used to synopsize data from a sample exercising the mean or standard deviation. Inferential statistics are used when data is viewed as a subclass of a specific population.

Types of Statistics

Statistics is a general, broad term, so it's natural that under that umbrella there exist a number of different models.

Mean

A mean is the mathematical average of a group of two or more numerals. The mean for a specified set of numbers can be computed in multiple ways, including the arithmetic mean, which shows how well a specific commodity performs over time, and the geometric mean, which shows the performance results of an investor's portfolio invested in that same commodity over the same period.

Regression Analysis

Regression analysis determines the extent to which specific factors such as interest rates, the price of a product or service, or particular industries or sectors influence the price fluctuations of an asset. This is depicted in the form of a straight line called linear regression.

Skewness

Skewness describes the degree a set of data varies from the standard distribution in a set of statistical data. Most data sets, including commodity returns and stock prices, have either positive skew, a curve skewed toward the left of the data average, or negative skew, a curve skewed toward the right of the data average.

Kurtosis

Kurtosis measures whether the data are light-tailed (less outlier-prone) or heavy-tailed (more outlier-prone) than the normal distribution. Data sets with high kurtosis have heavy tails, or outliers, which implies greater investment risk in the form of occasional wild returns. Data sets with low kurtosis have light tails, or lack of outliers, which implies lesser investment risk.

Variance

Variance is a measurement of the span of numbers in a data set. The variance measures the distance each number in the set is from the mean. Variance can help determine the risk an investor might accept when buying an investment.

Ronald Fisher developed the analysis of variance method. It is used to decide the effect solitary variables have on a variable that is dependent. It may be used to compare the performance of different stocks over time.

Statistics can be applied to various different problems and situations but the underlying concepts all remain the same. Thus it is important to understand what statistics is, not only from an application point of view but also from an interpretation point of view. This is required because of the diverse applications of statistics, from social science experiments to studying quantum mechanical phenomena.

Statistics can be broadly classified into descriptive statistics and inferential statistics.

To understand statistics, one needs to study and understand the probability theory. These are closely connected and inseparable in most cases. In fact, historically, the foundations of statistics were laid with the development of probability theory.

The ideas of presenting data and drawing relevant inferences are central to the successful use of statistical theory. In the end, the statistical analysis should be able to tell us something concrete about the sample that we are studying. A number of errors are possible in the interpretation of statistical results and a careful analysis needs to be made to prevent these errors.

In some rare cases, statistics can be used to draw conclusions that appear to be statistically relevant but on careful examination, are not. When such practices are intentional, they can be hard to detect. One good example of such statistical misconduct is data dredging. Therefore one should also be able to spot the scope and relevance of a statistical study and understand it in the context of the study within which it was intended.

CHARACTERISTICS OF STATISTICS

Some of the potential characteristics that a statistic should include:

Completeness

Completeness refers to an indication of whether or not the data required to meet the information demand is available in the data resource. Completeness of data is necessary to ensure the accuracy of the observed data.

Consistency

Consistency is viewed in terms of the uniformity or stability of data. Some of the statistics used to measure consistency include standard deviation, range, and variance. When measuring the consistency of data from a sample that is representative of a large population, the standard error of the mean is usually examined.

Also, when using instruments to collect data, the consistency can be measured by estimating the reliability of the obtained scores.

Sufficiency

A statistic is considered sufficient if there is no other statistic that can be computed from the sample. The sufficiency concept is common in descriptive statistics due to its strong dependence on the assumption of the data distribution form.

Unbiasedness

The bias of a statistics is determined by the difference between the true value of the parameter being measured and the estimator's expected value. If the mean of the sampling distribution and the expected value of the parameter are equal, the statistic is considered to be unbiased.

STATISTICAL DATA

When working with statistics, it's important to recognize the different types of data: numerical (discrete and continuous), categorical, and ordinal. Data are the actual pieces of information that you collect through your study. For example, if you ask five of your friends how many pets they own, they might give you the following data: 0, 2, 1, 4, 18. (The fifth friend might count each of her aquarium fish as a separate pet.) Not all data are numbers; let's say you also record the gender of each of your friends, getting the following data: male, male, female, male, female.

Most data fall into one of two groups: numerical or categorical:

- Numerical data: These data have meaning as a measurement, such as a person's height, weight, IQ, or blood pressure; or they're a count, such as the number of stock shares a person owns, how many teeth a dog has, or how many pages you can read of your favorite book before you fall asleep. (Statisticians also call numerical data quantitative data.). Numerical data can be further broken into two types: discrete and continuous:
  - Discrete data represent items that can be counted; they take on possible values that can be listed out. The list of possible values may be fixed (also called finite); or it may go from 0, 1, 2, on to infinity (making it countably infinite). For example, the number of heads in 100 coin flips takes on values from 0 through 100 (finite case), but the number of flips needed to get 100 heads takes on values from 100 (the fastest scenario) on up to infinity (if you never get to that 100th heads). Its possible values are listed as 100, 101, 102, 103, . . . (representing the countably infinite case).
  - Continuous data represent measurements; their possible values cannot be counted and can only be described using intervals on the real number line.

For example, the exact amount of gas purchased at the pump for cars with 20-gallon tanks would be continuous data from 0 gallons to 20 gallons, represented by the interval [0, 20], inclusive. You might pump 8.40 gallons, or 8.41, or 8.414863 gallons, or any possible number from 0 to 20. In this way, continuous data can be thought of as being uncountably infinite. For ease of recordkeeping, statisticians usually pick some point in the number to round off. Another example would be that the lifetime of a C battery can be anywhere from 0 hours to an infinite number of hours (if it lasts forever), technically, with all possible values in between. Granted, you don't expect a battery to last more than a few hundred hours, but no one can put a cap on how long it can go.

- Categorical data: Categorical data represent characteristics such as a person's gender, marital status, hometown, or the types of movies they like. Categorical data can take on numerical values (such as "1" indicating male and "2" indicating female), but those numbers don't have mathematical meaning. You couldn't add them together, for example. (Other names for categorical data are qualitative data, or Yes/No data.)

Ordinal data mixes numerical and categorical data. The data fall into categories, but the numbers placed on the categories have meaning. For example, rating a restaurant on a scale from 0 (lowest) to 4 (highest) stars gives ordinal data. Ordinal data are often treated as categorical, where the groups are ordered when graphs and charts are made. However, unlike categorical data, the numbers do have mathematical meaning. For example, if you survey 100 people and ask them to rate a restaurant on a scale from 0 to 4, taking the average of the 100 responses will have meaning. This would not be the case with categorical data.

STATISTICAL DATA ANALYSIS

Statistics is basically a science that involves data collection, data interpretation and finally, data validation. Statistical data analysis is a procedure of performing various statistical operations. It is a kind of quantitative research, which seeks to quantify the data, and typically, applies some form of statistical analysis. Quantitative data basically involves descriptive data, such as survey data and observational data.

Statistical data analysis generally involves some form of statistical tools, which a layman cannot perform without having any statistical knowledge. There are various software packages to perform statistical data analysis. This software includes Statistical Analysis System (SAS), Statistical Package for the Social Sciences (SPSS), Stat soft, etc.

Data in statistical data analysis consists of variables. Sometimes the data is univariate or multivariate. Depending upon the number of variables, the researcher performs different statistical techniques.

If the data in statistical data analysis is multiple in numbers, then several multivariates can be performed. These are factor statistical data analysis, discriminant statistical data analysis, etc. Similarly, if the data is singular in number, then the univariate statistical data analysis is performed. This includes t test for significance, z test, f test, ANOVA one way, etc.

The data in statistical data analysis is basically of 2 types, namely, continuous data and discreet data. The continuous data is the one that cannot be counted. For example, intensity of a light can be measured but cannot be counted. The discreet data is the one that can be counted. For example, the number of bulbs can be counted.

The continuous data in statistical data analysis is distributed under continuous distribution function, which can also be called the probability density function, or simply pdf.

The discreet data in statistical data analysis is distributed under discreet distribution function, which can also be called the probability mass function or simple pmf.

We use the word 'density' in continuous data of statistical data analysis because density cannot be counted, but can be measured. We use the word 'mass' in discreet data of statistical data analysis because mass cannot be counted.

There are various pdf's and pmf's in statistical data analysis. For example, Poisson distribution is the commonly known pmf, and normal distribution is the commonly known pdf.

These distributions in statistical data analysis help us to understand which data falls under which distribution. If the data is about the intensity of a bulb, then the data would be falling in Poisson distribution.

There is a major task in statistical data analysis, which comprises of statistical inference. The statistical inference is mainly comprised of two parts: Estimation and tests of hypothesis.

Estimation in statistical data analysis mainly involves parametric data—the data that consists of parameters. On the other hand, tests of hypothesis in statistical data analysis mainly involve non parametric data— the data that consists of no parameters.

Methods for Statistical Data Analysis

Mean

The arithmetic mean, more commonly known as "the average," is the sum of a list of numbers divided by the number of items on the list. The mean is useful in determining the overall trend of a data set or providing a rapid snapshot of your data. Another advantage of the mean is that it's very easy and quick to calculate.

Pitfall:

Taken alone, the mean is a dangerous tool. In some data sets, the mean is also closely related to the mode and the median (two other measurements near the average). However, in a data set with a high number of outliers or a skewed distribution, the mean simply doesn't provide the accuracy you need for a nuanced decision.

Standard Deviation

The standard deviation, often represented with the Greek letter sigma, is the measure of a spread of data around the mean. A high standard deviation signifies that data is spread more widely from the mean, where a low standard deviation signals that more data align with the mean. In a portfolio of data analysis methods, the standard deviation is useful for quickly determining dispersion of data points.

Pitfall:

The standard deviation is deceptive if taken alone. For example, if the data have a very strange pattern such as a non-normal curve or a large amount of outliers, then the standard deviation won't give you all the information you need.

Regression

Regression models the relationships between dependent and explanatory variables, which are usually charted on a scatterplot. The regression line also designates whether those relationships are strong or weak. Regression is commonly taught in high school or college statistics courses with applications for science or business in determining trends over time.

Pitfall:

Regression is not very nuanced. Sometimes, the outliers on a scatterplot (and the reasons for them) matter significantly. For example, an outlying data point may represent the input from your most critical supplier or your highest selling product. The nature of a regression line, however, tempts you to ignore these outliers. As an illustration, examine a picture of ANSCOMBE'S QUARTET, in which the data sets have the exact same regression line but include widely different data points.

Sample Size Determination

When measuring a large data set or population, like a workforce, you don't always need to collect information from every member of that population – a sample does the job just as well. The trick is to determine the right size for a sample to be accurate. Using proportion and standard deviation methods, you are able to accurately determine the right sample size you need to make your data collection statistically significant.

Pitfall:

When studying a new, untested variable in a population, your proportion equations might need to rely on certain assumptions. However, these assumptions might be completely inaccurate. This error is then passed along to your sample size determination and then onto the rest of your statistical data analysis

Hypothesis Testing

Also commonly called t testing, hypothesis testing assesses if a certain premise is actually true for your data set or population. In data analysis and statistics, you consider the result of a hypothesis test statistically significant if the results couldn't have happened by random chance. Hypothesis tests are used in everything from science and research to business and economic.

Pitfall:

To be rigorous, hypothesis tests need to watch out for common errors. For example, the placebo effect occurs when participants falsely expect a certain result and then perceive (or actually attain) that result. Another common error is the Hawthorne effect (or observer effect), which happens when participants skew results because they know they are being studied.

Overall, these methods of data analysis add a lot of insight to your DECISION-MAKING PORTFOLIO, particularly if you've never analyzed a process or data set with statistics before.

References

- Statistics: investopedia.com, Retrieved 02 July, 2019
- What-is-statistics: explorable.com, Retrieved 10 March, 2019
- Types-of-statistical-data-numerical-categorical-and-ordinal', math-statistics: dummies.com, Retrieved 25 May, 2019
- Statistical-data-analysis: statisticssolutions.com, Retrieved 28 January, 2019
- 5-Most-Important-Methods-For-Statistical-Data-Analysis- 356764: bigskyassociates.com, Retrieved 18 April, 2019

2

Branches of Statistics

The discipline of statistics can be categorized into various branches such as descriptive analysis, parametric and nonparametric statistics, exact statistics, etc. This chapter closely examines these branches of statistics to provide an extensive understanding of the subject.

DESCRIPTIVE STATISTICS

Descriptive statistics are brief descriptive coefficients that summarize a given data set, which can be either a representation of the entire or a sample of a population. Descriptive statistics are broken down into measures of central tendency and measures of variability (spread). Measures of central tendency include the mean, median, and mode, while measures of variability include the standard deviation, variance, the minimum and maximum variables, and the kurtosis and skewness.

Descriptive statistics, in short, help describe and understand the features of a specific data set by giving short summaries about the sample and measures of the data. The most recognized types of descriptive statistics are measures of center: the mean, median, and mode, which are used at almost all levels of math and statistics. The mean, or the average, is calculated by adding all the figures within the data set and then dividing by the number of figures within the set. For example, the sum of the following data set is 20: (2, 3, 4, 5, 6). The mean is 4 (20/5). The mode of a data set is the value appearing most often, and the median is the figure situated in the middle of the data set. It is the figure separating the higher figures from the lower figures within a data set. However, there are less-common types of descriptive statistics that are still very important.

People use descriptive statistics to repurpose hard-to-understand quantitative insights across a large data set into bite-sized descriptions. A student's grade point average (GPA), for example, provides a good understanding of descriptive statistics. The idea of a GPA is that it takes data points from a wide range of exams, classes, and grades, and averages them together to provide a general understanding of a student's overall academic abilities. A student's personal GPA reflects his mean academic performance.

Measures of Descriptive Statistics

All descriptive statistics are either measures of central tendency or measures of variability, also known as measures of dispersion. Measures of central tendency focus on the average or middle values of data sets; whereas, measures of variability focus on the dispersion of data. These two measures use graphs, tables, and general discussions to help people understand the meaning of the analyzed data.

Measures of central tendency describe the center position of a distribution for a data set. A person analyzes the frequency of each data point in the distribution and describes it using the mean, median, or mode, which measures the most common patterns of the analyzed data set.

Measures of variability, or the measures of spread, aid in analyzing how spread-out the distribution is for a set of data. For example, while the measures of central tendency may give a person the average of a data set, it does not describe how the data is distributed within the set. So, while the average of the data may be 65 out of 100, there can still be data points at both 1 and 100. Measures of variability help communicate this by describing the shape and spread of the data set. Range, quartiles, absolute deviation, and variance are all examples of measures of variability. Consider the following data set: 5, 19, 24, 62, 91, 100. The range of that data set is 95, which is calculated by subtracting the lowest number (5) in the data set from the highest (100).

PARAMETRIC STATISTICS

Parametric statistics is a branch of statistics which assumes that sample data come from a population that can be adequately modeled by a probability distribution that has a fixed set of parameters. Conversely a non-parametric model differs precisely in that the parameter set (or feature set in machine learning) is not fixed and can increase, or even decrease, if new relevant information is collected.

Most well-known statistical methods are parametric. Regarding nonparametric (and semiparametric) models, Sir David Cox has said, “These typically involve fewer assumptions of structure and distributional form but usually contain strong assumptions about independencies”.

The normal family of distributions all have the same general shape and are *parameterized* by mean and standard deviation. That means that if the mean and standard deviation are known and if the distribution is normal, the probability of any future observation lying in a given range is known.

Suppose that we have a sample of 99 test scores with a mean of 100 and a standard deviation of 1. If we assume all 99 test scores are random observations from a normal distribution, then we predict there is a 1% chance that the 100th test score will be higher than 102.33 (that is, the mean plus 2.33 standard deviations), assuming that the 100th test score comes from the same distribution as the others. Parametric statistical methods are used to compute the 2.33 value above, given 99 independent observations from the same normal distribution.

A non-parametric estimate of the same thing is the maximum of the first 99 scores. We don't need to assume anything about the distribution of test scores to reason that before we gave the test it was equally likely that the highest score would be any of the first 100. Thus there is a 1% chance that the 100th score is higher than any of the 99 that preceded it.

EXACT STATISTICS

Exact statistics, such as that described in exact test, is a branch of statistics that was developed to provide more accurate results pertaining to statistical testing and interval estimation by eliminating procedures based on asymptotic and approximate statistical methods. The main characteristic of exact methods is that statistical tests and confidence intervals are based on exact probability statements that are valid for any sample size. Exact statistical methods help avoid some of the unreasonable assumptions of traditional statistical methods, such as the assumption of equal variances in classical ANOVA. They also allow exact inference on variance components of mixed models.

When exact p-values and confidence intervals are computed under a certain distribution, such as the normal distribution, then the underlying methods are referred to as exact parametric methods. The exact methods that do not make any distributional assumptions are referred to as exact nonparametric methods. The latter has the advantage of making fewer assumptions whereas, the former tend to yield more powerful tests when the distributional assumption is reasonable. For advanced methods such as higher-way ANOVA regression analysis, and mixed models, only exact parametric methods are available.

When the sample size is small, asymptotic results given by some traditional methods may not be valid. In such situations, the asymptotic p-values may differ substantially from the exact p-values. Hence asymptotic and other approximate results may lead to unreliable and misleading conclusions.

Approach

All classical statistical procedures are constructed using statistics which depend only on observable random vectors, whereas generalized estimators, tests, and confidence intervals used in exact statistics take advantage of the observable random vectors and the observed values both, as in the Bayesian approach but without having to treat constant parameters as random variables. For example, in sampling from a normal population with mean μ and variance σ^2, suppose $\overline{X}$ and S^2 are the sample mean and the sample variance. Then, defining Z and U thus:

$$Z = \sqrt{n}(\overline{X} - \mu) / \sigma \sim N(0,1)$$

and that,

$$U = nS^2 / \sigma^2 \sim \chi^2_{n-1}.$$

Now suppose the parameter of interest is the coefficient of variation, $\rho = \mu / \sigma$. then, we can easily perform exact tests and exact confidence intervals for ρ based on the generalized statistic:

$$R = \frac{\overline{x}S}{s\sigma} - \frac{\overline{X} - \mu}{\sigma} = \frac{\overline{x}}{s}\frac{\sqrt{U}}{\sqrt{n}} - \frac{Z}{\sqrt{n}},$$

Where, $\overline{x}$ is the observed value of $\overline{X}$ and S is the observed value of s. Exact inferences on ρ based on probabilities and expected values of R are possible because its distribution and the observed value are both free of nuisance parameters.

Generalized p-values

Classical statistical methods do not provide exact tests to many statistical problems such as testing Variance Components and ANOVA under unequal variances. To rectify this situation, the generalized *p*-values are defined as an extension of the classical *p*-values so that one can perform tests based on exact probability statements valid for any sample size.

NONPARAMETRIC STATISTICS

Nonparametric statistics is the branch of statistics that is not based solely on parametrized families of probability distributions (common examples of parameters are the mean and variance). Nonparametric statistics is based on either being distribution-free or having a specified distribution but with the distribution's parameters unspecified. Nonparametric statistics includes both descriptive statistics and statistical inference.

The term "nonparametric statistics" has been imprecisely defined in the following two ways, among others.

- The first meaning of *nonparametric* covers techniques that do not rely on data belonging to any particular parametric family of probability distributions. These include, among others:

- Distribution free methods, which do not rely on assumptions that the data are drawn from a given parametric family of probability distributions. As such it is the opposite of parametric statistics.
- Nonparametric statistics (a statistic is defined to be a function on a sample; no dependency on a parameter).

Order statistics, which are based on the ranks of observations, is one example of such statistics.

Statistical hypotheses concern the behavior of observable random variables. For example, the hypothesis (a) that a normal distribution has a specified mean and variance is statistical; so is the hypothesis (b) that it has a given mean but unspecified variance; so is the hypothesis (c) that a distribution is of normal form with both mean and variance unspecified; finally, so is the hypothesis (d) that two unspecified continuous distributions are identical.

It will have been noticed that in the examples (a) and (b) the distribution underlying the observations was taken to be of a certain form (the normal) and the hypothesis was concerned entirely with the value of one or both of its parameters. Such a hypothesis, for obvious reasons, is called parametric.

Hypothesis (c) was of a different nature, as no parameter values are specified in the statement of the hypothesis; we might reasonably call such a hypothesis non-parametric. Hypothesis (d) is also non-parametric but, in addition, it does not even specify the underlying form of the distribution and may now be reasonably termed distribution-free. Notwithstanding these distinctions, the statistical literature now commonly applies the label "non-parametric" to test procedures that we have just termed "distribution-free", thereby losing a useful classification.

- The second meaning of non-parametric covers techniques that do not assume that the structure of a model is fixed. Typically, the model grows in size to accommodate the complexity of the data. In these techniques, individual variables are typically assumed to belong to parametric distributions, and assumptions about the types of connections among variables are also made. These techniques include, among others:
  - Non-parametric regression, which is modeling whereby the structure of the relationship between variables is treated non-parametrically, but where nevertheless there may be parametric assumptions about the distribution of model residuals.
  - Non-parametric hierarchical Bayesian models, such as models based on the Dirichlet process, which allow the number of latent variables to grow as

necessary to fit the data, but where individual variables still follow parametric distributions and even the process controlling the rate of growth of latent variables follows a parametric distribution.

Applications and Purpose

Non-parametric methods are widely used for studying populations that take on a ranked order (such as movie reviews receiving one to four stars). The use of non-parametric methods may be necessary when data have a ranking but no clear numerical interpretation, such as when assessing preferences. In terms of levels of measurement, non-parametric methods result in ordinal data.

As non-parametric methods make fewer assumptions, their applicability is much wider than the corresponding parametric methods. In particular, they may be applied in situations where less is known about the application in question. Also, due to the reliance on fewer assumptions, non-parametric methods are more robust.

Another justification for the use of non-parametric methods is simplicity. In certain cases, even when the use of parametric methods is justified, non-parametric methods may be easier to use. Due both to this simplicity and to their greater robustness, non-parametric methods are seen by some statisticians as leaving less room for improper use and misunderstanding.

The wider applicability and increased robustness of non-parametric tests comes at a cost: In cases where a parametric test would be appropriate, non-parametric tests have less power. In other words, a larger sample size can be required to draw conclusions with the same degree of confidence.

Non-parametric Models

Non-parametric models differ from parametric models in that the model structure is not specified a priori but is instead determined from data. The term non-parametric is not meant to imply that such models completely lack parameters but that the number and nature of the parameters are flexible and not fixed in advance:

- A histogram is a simple nonparametric estimate of a probability distribution.
- Kernel density estimation provides better estimates of the density than histograms.
- Nonparametric regression and semiparametric regression methods have been developed based on kernels, splines, and wavelets.
- Data envelopment analysis provides efficiency coefficients similar to those obtained by multivariate analysis without any distributional assumption.
- KNNs classify the unseen instance based on the K points in the training set which are nearest to it.

- A support vector machine (with a Gaussian kernel) is a nonparametric large-margin classifier.
- Method of moments (statistics) with polynomial probability distributions.

Methods

Non-parametric (or distribution-free) inferential statistical methods are mathematical procedures for statistical hypothesis testing which, unlike parametric statistics, make no assumptions about the probability distributions of the variables being assessed. The most frequently used tests include:

- Analysis of similarities.
- Anderson–Darling test: Tests whether a sample is drawn from a given distribution.
- Statistical bootstrap methods: Estimates the accuracy/sampling distribution of a statistic.
- Cochran's Q: Tests whether *k* treatments in randomized block designs with 0/1 outcomes have identical effects.
- Cohen's kappa: Measures inter-rater agreement for categorical items.
- Friedman two-way analysis of variance by ranks: tests whether *k* treatments in randomized block designs have identical effects.
- Kaplan–Meier: Estimates the survival function from lifetime data, modeling censoring.
- Kendall's tau: Measures statistical dependence between two variables.
- Kendall's W: A measure between 0 and 1 of inter-rater agreement.
- Kolmogorov–Smirnov test: Tests whether a sample is drawn from a given distribution, or whether two samples are drawn from the same distribution.
- Kruskal–Wallis one-way analysis of variance by ranks: Tests whether > 2 independent samples are drawn from the same distribution.
- Kuiper's test: Tests whether a sample is drawn from a given distribution, sensitive to cyclic variations such as day of the week.
- Logrank test: Compares survival distributions of two right-skewed, censored samples.
- Mann–Whitney U or Wilcoxon rank sum test: Tests whether two samples are drawn from the same distribution, as compared to a given alternative hypothesis.

- McNemar's test: Tests whether, in 2×2 contingency tables with a dichotomous trait and matched pairs of subjects, row and column marginal frequencies are equal.
- Median test: Tests whether two samples are drawn from distributions with equal medians.
- Pitman's permutation test: A statistical significance test that yields exact p values by examining all possible rearrangements of labels.
- Rank products: Detects differentially expressed genes in replicated microarray experiments.
- Siegel–Tukey test: Tests for differences in scale between two groups.
- Sign test: Tests whether matched pair samples are drawn from distributions with equal medians.
- Spearman's rank correlation coefficient: Measures statistical dependence between two variables using a monotonic function.
- Squared ranks test: Tests equality of variances in two or more samples.
- Tukey–Duckworth test: Tests equality of two distributions by using ranks.
- Wald–Wolfowitz runs test: Tests whether the elements of a sequence are mutually independent/random.
- Wilcoxon signed-rank test: Tests whether matched pair samples are drawn from populations with different mean ranks.

ESTIMATION THEORY

Estimation theory is a branch of statistics that deals with estimating the values of parameters based on measured empirical data that has a random component. The parameters describe an underlying physical setting in such a way that their value affects the distribution of the measured data. An estimator attempts to approximate the unknown parameters using the measurements.

In estimation theory, two approaches are generally considered.

- The probabilistic approach assumes that the measured data is random with probability distribution dependent on the parameters of interest.
- The set-membership approach assumes that the measured data vector belongs to a set which depends on the parameter vector.

Examples:

It is desired to estimate the proportion of a population of voters who will vote for a particular candidate. That proportion is the parameter sought; the estimate is based on a small random sample of voters. Alternatively, it is desired to estimate the probability of a voter voting for a particular candidate, based on some demographic features, such as age.

Or, for example, in radar the aim is to find the range of objects (airplanes, boats, etc.) by analyzing the two-way transit timing of received echoes of transmitted pulses. Since the reflected pulses are unavoidably embedded in electrical noise, their measured values are randomly distributed, so that the transit time must be estimated.

As another example, in electrical communication theory, the measurements which contain information regarding the parameters of interest are often associated with a noisy signal.

Basics

For a given model, several statistical "ingredients" are needed so the estimator can be implemented. The first is a statistical sample – a set of data points taken from a random vector (RV) of size N. Put into a vector:

$$\mathrm{x} = \begin{bmatrix} x[0] \\ x[1] \\ \vdots \\ x[N-1] \end{bmatrix}.$$

Secondly, there are M parameters:

$$\theta = \begin{bmatrix} \theta_1 \\ \theta_2 \\ \vdots \\ \theta_M \end{bmatrix},$$

Whose, values are to be estimated. Third, the continuous probability density function (pdf) or its discrete counterpart, the probability mass function (pmf), of the underlying distribution that generated the data must be stated conditional on the values of the parameters:

$$p(\mathrm{x} \mid \theta).$$

It is also possible for the parameters themselves to have a probability distribution (e.g., Bayesian statistics). It is then necessary to define the Bayesian probability:

$$\pi(\theta).$$

After the model is formed, the goal is to estimate the parameters, with the estimates commonly denoted $\hat{\theta}$, where the "hat" indicates the estimate.

One common estimator is the minimum mean squared error (MMSE) estimator, which utilizes the error between the estimated parameters and the actual value of the parameters:

$$\mathrm{e} = \hat{\theta} - \theta$$

As the basis for optimality. This error term is then squared and the expected value of this squared value is minimized for the MMSE estimator.

Estimators

Commonly used estimators (estimation methods) and topics related to them include:

- Maximum likelihood estimators.
- Bayes estimators.
- Method of moments estimators.
- Cramér–Rao bound.
- Least squares.
- Minimum mean squared error (MMSE), also known as Bayes least squared error (BLSE).
- Maximum a posteriori (MAP).
- Minimum variance unbiased estimator (MVUE).
- Nonlinear system identification.
- Best linear unbiased estimator (BLUE).
- Unbiased estimators.
- Particle filter.
- Markov chain Monte Carlo (MCMC).
- Kalman filter, and its various derivatives.
- Wiener filter.

Unknown Constant in Additive white Gaussian Noise

Consider a received discrete signal, $x[n]$, of N independent samples that consists of an unknown constant A with additive white Gaussian noise (AWGN) $w[n]$ with known variance σ (*i.e.*, $\mathcal{N}(0,\sigma^2)$). Since the variance is known then the only unknown parameter is A.

The model for the signal is then,

$$x[n] = A + w[n] \quad n = 0,1,\ldots,N-1$$

Two possible (of many) estimators for the parameter A are:

- $\hat{A}_1 = x[0]$
- $\hat{A}_2 = \frac{1}{N}\sum_{n=0}^{N-1} x[n]$

Both of these estimators have a mean of A, which can be shown through taking the expected value of each estimator:

$$\mathrm{E}\left[\hat{A}_1\right] = \mathrm{E}\left[x[0]\right] = A$$

and

$$\mathrm{E}\left[\hat{A}_2\right] = \mathrm{E}\left[\frac{1}{N}\sum_{n=0}^{N-1} x[n]\right] = \frac{1}{N}\left[\sum_{n=0}^{N-1}\mathrm{E}\left[x[n]\right]\right] = \frac{1}{N}[NA] = A$$

At this point, these two estimators would appear to perform the same. However, the difference between them becomes apparent when comparing the variances.

$$\mathrm{var}\left(\hat{A}_1\right) = \mathrm{var}\left(x[0]\right) = \sigma^2$$

and

$$\mathrm{var}(\hat{A}_2) = \mathrm{var}\left(\frac{1}{N}\sum_{n=0}^{N-1} x[n]\right) \overset{\text{independence}}{=} \frac{1}{N^2}\left[\sum_{n=0}^{N-1}\mathrm{var}(x[n])\right] = \frac{1}{N^2}\left[N\sigma^2\right] = \frac{\sigma^2}{N}$$

It would seem that the sample mean is a better estimator since its variance is lower for every $N > 1$.

Maximum Likelihood

Continuing the example using he maximum likelihood estimator, the probability density function (pdf) of the noise for one sample $w[n]$ is:

$$p(w[n]) = \frac{1}{\sigma\sqrt{2\pi}}\exp\left(-\frac{1}{2\sigma^2}w[n]^2\right)$$

and the probability of $x[n]$ becomes $x[n]$ can be thought of a $\mathcal{N}(A,\sigma^2)$)

$$p(x[n];A) = \frac{1}{\sigma\sqrt{2\pi}}\exp\left(-\frac{1}{2\sigma^2}(x[n]-A)^2\right)$$

By independence, the probability of x becomes,

$$p(x;A) = \prod_{n=0}^{N-1} p\,(x[n];A) = \frac{1}{(\sigma\sqrt{2\pi})^N}\exp\left(-\frac{1}{2\sigma^2}\sum_{n=0}^{N-1}(x[n]-A)^2\right)$$

Taking the natural logarithm of the pdf:

$$\ln p(\mathrm{x};A) = -N\ln\left(\sigma\sqrt{2\pi}\right) - \frac{1}{2\sigma^2}\sum_{n=0}^{N-1}(x[n]-A)^2$$

and the maximum likelihood estimator is:

$$\hat{A} \quad \arg\max \ln p(\mathrm{x};A)$$

Taking the first derivative of the log-likelihood function:

$$\frac{\partial}{\partial A}\ln p(\mathrm{x};A) = \frac{1}{\sigma^2}\left[\sum_{n=0}^{N-1}(x[n]-A)\right] = \frac{1}{\sigma^2}\left[\sum_{n=0}^{N-1}x[n]-NA\right]$$

and setting it to zero:

$$0 = \frac{1}{\sigma^2}\left[\sum_{n=0}^{N-1}x[n]-NA\right] = \sum_{n=0}^{N-1}x[n]-NA$$

This results in the maximum likelihood estimator:

$$\hat{A} = \frac{1}{N}\sum_{n=0}^{N-1}x[n]$$

Which is simply the sample mean? From this example, it was found that the sample mean is the maximum likelihood estimator for N samples of a fixed, unknown parameter corrupted by AWGN.

Cramér–Rao Lower Bound

To find the Cramér–Rao lower bound (CRLB) of the sample mean estimator, it is first necessary to find the Fisher information number:

$$\mathcal{I}(A) = \mathrm{E}\left(\left[\frac{\partial}{\partial A}\ln p(\mathrm{x};A)\right]^2\right) = -\mathrm{E}\left[\frac{\partial^2}{\partial A^2}\ln p(\mathrm{x};A)\right]$$

and copying from above,

$$\frac{\partial}{\partial A}\ln p(\mathrm{x};A)=\frac{1}{\sigma^2}\left[\sum_{n=0}^{N-1}x[n]-NA\right]$$

Taking the second derivative:

$$\frac{\partial^2}{\partial A^2}\ln p(\mathrm{x};A)=\frac{1}{\sigma^2}(-N)=\frac{-N}{\sigma^2}$$

and finding the negative expected value is trivial since it is now a deterministic constant:

$$-\mathrm{E}\left[\frac{\partial^2}{\partial A^2}\ln p(\mathrm{x};A)\right]=\frac{N}{\sigma^2}$$

Finally, putting the Fisher information into:

$$\mathrm{var}\left(\hat{A}\right)\geq\frac{1}{\mathcal{I}}$$

results in, $\mathrm{var}\left(\hat{A}\right)\geq\frac{\sigma^2}{N}$

Comparing this to the variance of the sample mean shows that the sample mean is equal to the Cramér–Rao lower bound for all values of N and A. In other words, the sample mean is the (necessarily unique) efficient estimator, and thus also the minimum variance unbiased estimator (MVUE), in addition to being the maximum likelihood estimator.

Maximum of a Uniform Distribution

One of the simplest non-trivial examples of estimation is the estimation of the maximum of a uniform distribution. It is used as a hands-on classroom exercise and to illustrate basic principles of estimation theory. Further, in the case of estimation based on a single sample, it demonstrates philosophical issues and possible misunderstandings in the use of maximum likelihood estimators and likelihood functions.

Given a discrete uniform distribution $1,2,\ldots,N$ with unknown maximum, the UMVU estimator for the maximum is given by,

$$\frac{k+1}{k}m-1=m+\frac{m}{k}-1$$

Where, m is the sample maximum and k is the sample size, sampling without replacement. This problem is commonly known as the German tank problem, due to application of maximum estimation to estimates of German tank production during World War II.

The formula may be understood intuitively as;

"The sample maximum plus the average gap between observations in the sample".

The gap being added to compensate for the negative bias of the sample maximum as an estimator for the population maximum.

This has a variance of:

$$\frac{1}{k}\frac{(N-k)(N+1)}{(k+2)} \approx \frac{N^2}{k^2} \text{ for small samples } k \ll N$$

So, a standard deviation of approximately N/k, the (population) average size of a gap between samples; compare $\frac{m}{k}$ above.

The sample maximum is the maximum likelihood estimator for the population maximum.

Applications

Numerous fields require the use of estimation theory. Some of these fields include (but are by no means limited to):

- Interpretation of scientific experiments.
- Signal processing.
- Clinical trials.
- Opinion polls.
- Quality control.
- Telecommunications.
- Project management.
- Software engineering.
- Control theory (in particular Adaptive control).
- Network intrusion detection system.
- Orbit determination.

Measured data are likely to be subject to noise or uncertainty and it is through statistical probability that optimal solutions are sought to extract as much information from the data as possible.

ECONOMIC STATISTICS

Economic Statistics is the branch of statistical science that studies the quantitative aspect of economic processes and phenomena in the national economy in conjunction with their qualitative aspect. Unlike more specialized forms of statistics, which study economic processes in particular branches of the national economy, economic statistics studies the national economy as an integrated whole.

Marxist-Leninist political economy constitutes the theoretical foundation of economic statistics. Political economy investigates and identifies the most important features of social-production relations and reveals the laws governing the production and distribution of material goods; economic statistics makes use of these principles to provide a quantitative description of phenomena and processes in the national economy and shows, with the help of economic-statistical indicators, how social production is developing in a particular place and time. In V. I. Lenin's definition, the purpose of economic statistics is to give "statistical expression" to the phenomena and laws of socioeconomic development of society. A preliminary and comprehensive socioeconomic analysis of the phenomena under study is a major precondition for the scientific organization of economic statistics.

Economic statistics is both an integral part of statistical science and an important branch of practical activity. Although it emerged as an independent scientific discipline and a subject taught in educational institutions before the Great Patriotic War of 1941–45, economic statistics underwent its greatest development after the war. In the investigation of economic processes and in the collection, processing, and analysis of statistical data, it makes extensive use of such techniques as mass statistical observation, grouping, indexing, the analysis of time series, and the balance method. Mathematicoeconomic research methods involving computers are coming to be used more widely.

In a socialist society, economic statistics is an important tool in the management and planning of the national economy. It describes the condition and development of a socialist economy, progress toward the fulfillment of national economic plans, and the way the branches of the economy are developing in relation to one another. In addition, it provides a picture of the introduction of new technology, the location of productive forces in the country, and improvements in public welfare. The most important tasks of economic statistics include the economic-statistical description of the efficiency of social production and the improvement of performance at all levels of the national economy.

In bourgeois statistics, economic statistics does not exist as an independent scientific discipline for the integrated investigation of processes and phenomena of social reproduction. The statistical literature of capitalist countries treats economic statistics as the application of general methods of statistics and mathematical statistics to the measurement of economic phenomena.

The system of indexes in economic statistics comprehensively describes economic processes and phenomena. The crucial indexes—the comprehensive, general indexes of economic development and of the growth of the people's material prosperity—include the total social product, national income, real incomes of the population, nonproductive consumption, accumulation, national wealth and its constituent elements, and the social productivity of labor.

Statisticians dealing with particular branches of the economy-apply the general principles that have been developed by economic statistics in order to construct a system of economic indexes and a methodology for calculating them. A major division of economic statistics is the balance of the national economy of the USSR, which makes it possible to ascertain whether the economy is developing in a balanced manner.

References

- Descriptive-statistics: investopedia.com, Retrieved 27 June, 2019
- Sheskin, David J. (2003) Handbook of Parametric and Nonparametric Statistical Procedures. CRC Press. ISBN 1-58488-440-1
- Fundamentals of Statistical Signal Processing: Estimation Theory by Steven M. Kay. ISBN 0-13-345711-7
- Economic-Statistics: encyclopedia2.thefreedictionary.com, 14 March, 2019

3

Statistical Measures

Statistical measures refer to the individual quantitative variable values for the statistical units in a specific group. Such measures include statistical mean, mode, median, range, skewness, quantile, quartile, variance, quartile deviation, pooled variance, standard deviation, etc. The topics elaborated in this chapter will help in gaining a better perspective of these statistical measures.

STATISTICAL MEAN

The statistical mean refers to the mean or average that is used to derive the central tendency of the data in question. It is determined by adding all the data points in a population and then dividing the total by the number of points. The resulting number is known as the mean or the average.

In mathematics and statistics, the term arithmetic mean is preferred over simply "mean" because it helps to differentiate between other means such as geometric and harmonic mean. Statistical mean is the most common term for calculating the mean of a statistical distribution.

An arithmetic mean is calculated using the following equation:

$$A := \frac{1}{n}\sum_{i=1}^{n} a_i$$

The statistical mean has a wide range of applicability in various types of experimentation. This type of calculation eliminates random errors and helps to derive a more accurate result than a result derived from a single experiment.

The statistical mean can also be used to interpret statistical data. Some important properties make statistical mean very useful for measuring central tendency. They are follows:

If numbers have average X, then:

Since Xi - X is the distance from a given number to the average. The numbers to the left of the mean are balanced by the numbers to the right of the mean. The residuals sum to zero only if a number is a statistical mean. A single number X is used as an estimate for the value of numbers, then the statistical mean minimizes the sum of the squares $(xi - X)^2$ of the residuals.

Statistical mean is popular because it includes every item in the data set and it can easily be used with other statistical measurements. However, the major disadvantage in using statistical mean is that it can be affected by extreme values in the data set and therefore be biased.

The statistical mean is widely used not only in the fields of mathematics and statistics, but also in economics, sociology and history. It gives important information about a data set and provides insight into the experiment and nature of the data.

The other terms used to measure central tendency (an average) are median and mode. In a normal distribution the statistical mean is equal to median and mode.

Arithmetic Mean

Arithmetic Mean in the most common and easily understood measure of central tendency. We can define mean as the value obtained by dividing the sum of measurements with the number of measurements contained in the data set and is denoted by the symbol $\overline{x}$.

Individual Data Series

When data is given on individual basis. Following is an example of individual series:

Items	5	10	20	30	40	50	60	70

Discrete Data Series

When data is given alongwith their frequencies. Following is an example of discrete series:

Items	5	10	20	30	40	50	60	70
Frequency	2	5	1	3	12	0	5	7

Continuous Data Series

When data is given based on ranges alongwith their frequencies. Following is an example of continous series:

Items	0-5	5-10	10-20	20-30	30-40
Frequency	2	5	1	3	12

MEDIAN

1, 3, 3, **6**, 7, 8, 9

Median = 6

1, 2, 3, **4**, **5**, 6, 8, 9

Median = (4 + 5) ÷ 2

= 4.5

Finding the median in sets of data with an odd and even number of values.

The median is the value separating the higher half from the lower half of a data sample (a population or a probability distribution). For a data set, it may be thought of as the "middle" value. For example, in the data set [1, 3, 3, 6, 7, 8, 9], the median is 6, the fourth largest, and also the fourth smallest, number in the sample. For a continuous probability distribution, the median is the value such that a number is equally likely to fall above or below it.

The median is a commonly used measure of the properties of a data set in statistics and probability theory. The basic advantage of the median in describing data compared to the mean (often simply described as the "average") is that it is not skewed so much by a small proportion of extremely large or small values, and so it may give a better idea of a "typical" value. For example, in understanding statistics like household income or assets, which vary greatly, the mean may be skewed by a small number of extremely high or low values. Median income, for example, may be a better way to suggest what a "typical" income is.

Because of this, the median is of central importance in robust statistics, as it is the most resistant statistic, having a breakdown point of 50%: so long as no more than half the data are contaminated, the median will not give an arbitrarily large or small result.

Finite Data Set of Numbers

The median of a finite list of numbers can be found by arranging all the numbers from smallest to greatest.

If there is an odd number of numbers, the middle one is picked. For example, consider the list of numbers

1, 3, 3, 6, 7, 8, 9

This list contains seven numbers. The median is the fourth of them, which is 6.

If there is an even number of observations, then there is no single middle value; the median is then usually defined to be the mean of the two middle values. For example, in the data set

1, 2, 3, 4, 5, 6, 8, 9

the median is the mean of the middle two numbers: this is $(4+5)/2,$, which is 4.5. (In more technical terms, this interprets the median as the fully trimmed mid-range).

The formula used to find the index of the middle number of a data set of n numerically ordered numbers is $(n+1)/2$. This either gives the middle number (for an odd number of values) or the halfway point between the two middle values. For example, with 14 values, the formula will give an index of 7.5, and the median will be taken by averaging the seventh (the floor of this index) and eighth (the ceiling of this index) values. So the median can be represented by the following formula:

$$\text{median}(a) = \frac{{}^{a}\lfloor(\#a+1)\div 2\rfloor + \lceil {}^{a}(\#a+1)\div 2\rceil}{2}$$

where a is an ordered list of numbers, $\#a$ denotes its length, and $\lfloor . \rfloor$ and $\lceil . \rceil$ denotes the floor and ceiling function, respectively.

Comparison of common averages of values [1, 2, 2, 3, 4, 7, 9]			
Type	Description	Example	Result
Arithmetic mean	Sum of values of a data set divided by number of values: $\overline{x} = \frac{1}{n}\sum_{i=1}^{n} x_i$	(1 + 2 + 2 + 3 + 4 + 7 + 9) / 7	4
Median	Middle value separating the greater and lesser halves of a data set	1, 2, 2, 3, 4, 7, 9	3
Mode	Most frequent value in a data set	1, 2, 2, 3, 4, 7, 9	2

One can find the median using the Stem-and-Leaf Plot.

There is no widely accepted standard notation for the median, but some authors represent the median of a variable x either as $\tilde{x}$ or as $\mu_{1/2}$ sometimes also M. In any of these cases, the use of these or other symbols for the median needs to be explicitly defined when they are introduced.

The median is used primarily for skewed distributions, which it summarizes differently from the arithmetic mean. Consider the multiset { 1, 2, 2, 2, 3, 14 }. The median is 2 in this case, (as is the mode), and it might be seen as a better indication of central tendency (less susceptible to the exceptionally large value in data) than the arithmetic mean of 4.

The median is a popular summary statistic used in descriptive statistics, since it is simple to understand and easy to calculate, while also giving a measure that is more robust in the presence of outlier values than is the mean. The widely cited empirical relationship between the relative locations of the mean and the median for skewed distributions is, however, not generally true. There are, however, various relationships for the *absolute* difference between them.

With an even number of observations no value need be exactly at the value of the median. Nonetheless, the value of the median is uniquely determined with the usual definition. A related concept, in which the outcome is forced to correspond to a member of the sample, is the medoid.

In a population, at most half have values strictly less than the median and at most half have values strictly greater than it. If each set contains less than half the population, then some of the population is exactly equal to the median. For example, if $a < b < c$, then the median of the list $\{a, b, c\}$ is b, and, if $a < b < c < d$, then the median of the list $\{a, b, c, d\}$ is the mean of b and c; i.e., it is $(b + c)/2$. As a median is based on the middle data in a set, it is not necessary to know the value of extreme results in order to calculate it. For example, in a psychology test investigating the time needed to solve a problem, if a small number of people failed to solve the problem at all in the given time a median can still be calculated.

The median can be used as a measure of location when a distribution is skewed, when end-values are not known, or when one requires reduced importance to be attached to outliers, e.g., because they may be measurement errors.

A median is only defined on ordered one-dimensional data, and is independent of any distance metric. A geometric median, on the other hand, is defined in any number of dimensions.

The median is one of a number of ways of summarising the typical values associated with members of a statistical population; thus, it is a possible location parameter. The median is the 2nd quartile, 5th decile, and 50th percentile. A median can be worked out for ranked but not numerical classes (e.g. working out a median grade when students are graded from A to F), although the result might be halfway between grades if there is an even number of cases.

When the median is used as a location parameter in descriptive statistics, there are several choices for a measure of variability: the range, the interquartile range, the mean absolute deviation, and the median absolute deviation.

For practical purposes, different measures of location and dispersion are often compared on the basis of how well the corresponding population values can be estimated from a sample of data. The median, estimated using the sample median, has good properties in this regard. While it is not usually optimal if a given population distribution is assumed, its properties are always reasonably good. For example, a comparison of the efficiency of candidate estimators shows that the sample mean is more statistically efficient than the sample median when data are uncontaminated by data from heavy-tailed distributions or from mixtures of distributions, but less efficient otherwise, and that the efficiency of the sample median is higher than that for a wide range of distributions. More specifically, the median has a 64% efficiency compared to the minimum-variance mean (for large normal samples), which is to say the variance of the median will be ~50% greater than the variance of the mean.

Aho give a divide-and-conquer algorithm to compute the kth smallest element of an unordered list a in linear time, which is faster than sorting. Running it with $k = \left\lceil \frac{\#a}{2} \right\rceil$ computes the median of a.

Probability Distributions

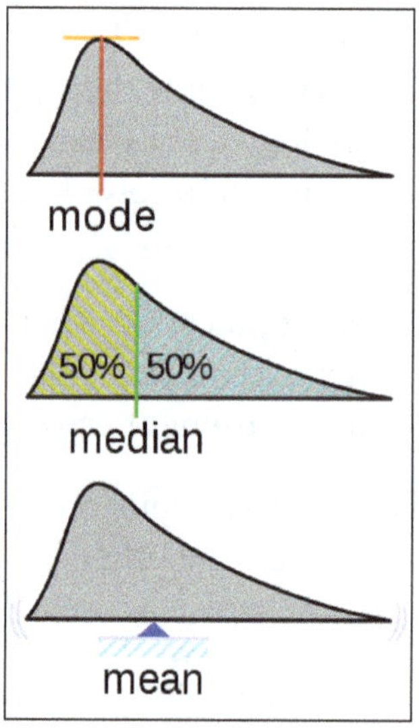

Geometric visualisation of the mode, median and mean of an arbitrary probability density function.

For any probability distribution on the real line R with cumulative distribution function *F*, regardless of whether it is any kind of continuous probability distribution, in particular an absolutely continuous distribution (which has a probability density function), or a discrete probability distribution, a median is by definition any real number m that satisfies the inequalities:

$$\mathrm{P}(X \leq m) \geq \frac{1}{2} \text{ and } \mathrm{P}(X \geq m) \geq \frac{1}{2}$$

or, equivalently, the inequalities,

$$\int_{(-\infty,m]} dF(x) \geq \frac{1}{2} \text{ and} \int_{[m,\infty)} dF(x) \geq \frac{1}{2}$$

in which a Lebesgue–Stieltjes integral is used. For an absolutely continuous probability distribution with probability density function f, the median satisfies,

$$P(X \geq m) = P(X \leq m) = \int_{-\infty}^{m} f(x)dx = \frac{1}{2}.$$

Any probability distribution on R has at least one median, but in specific cases there may be more than one median. Specifically, if a probability density is zero on an interval $[a, b]$, and the cumulative distribution function at a is 1/2, any value between a and b will also be a median.

Medians of Particular Distributions

The medians of certain types of distributions can be easily calculated from their parameters; furthermore, they exist even for some distributions lacking a well-defined mean, such as the Cauchy distribution:

- The median of a symmetric unimodal distribution coincides with the mode.
- The median of a symmetric distribution which possesses a mean μ also takes the value μ.
    - The median of a normal distribution with mean μ and variance σ^2 is μ. In fact, for a normal distribution, mean = median = mode.
    - The median of a uniform distribution in the interval $[a, b]$ is $(a + b) / 2$, which is also the mean.
- The median of a Cauchy distribution with location parameter x_0 and scale parameter y is x_0, the location parameter.
- The median of a power law distribution x^{-a}, with exponent $a > 1$ is $2^{1/(a-1)}x_{min}$, where x_{min} is the minimum value for which the power law holds.
- The median of an exponential distribution with rate parameter λ is the natural logarithm of 2 divided by the rate parameter: $\lambda^{-1}\ln 2$.
- The median of a Weibull distribution with shape parameter k and scale parameter λ is $\lambda(\ln 2)^{1/k}$.

Populations

Optimality Property

The *mean absolute error* of a real variable c with respect to the random variable X is,

$$E(|X - c|)$$

Provided that the probability distribution of X is such that the above expectation exists, then m is a median of X if and only if m is a minimizer of the mean absolute error with respect to X. In particular, m is a sample median if and only if m minimizes the arithmetic mean of the absolute deviations.

More generally, a median is defined as a minimum of

$$E(|X-c|-|X|),$$

This optimization-based definition of the median is useful in statistical data-analysis, for example, in k-medians clustering.

Unimodal Distributions

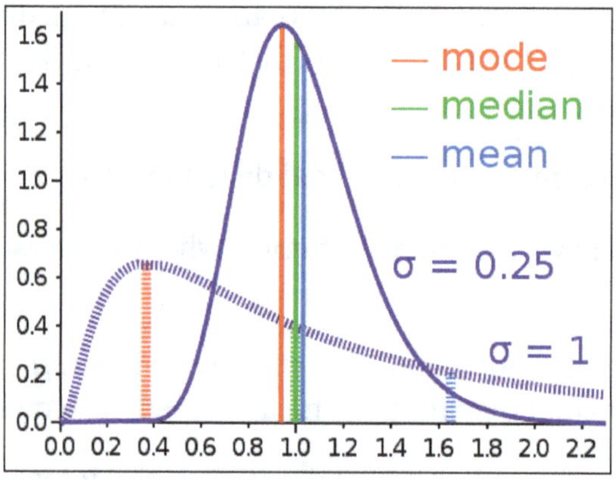

Comparison of mean, median and mode of two log-normal distributions with different skewness.

It can be shown for a unimodal distribution that the median $\tilde{X}$ and the mean $\bar{X}$ lie within $(3/5)^{1/2} \approx 0.7746$ standard deviations of each other. In symbols,

$$\frac{\tilde{X}-\bar{X}|}{\sigma} \leq \left(\frac{3}{5}\right)^{1/2}$$

where $|\cdot|$ is the absolute value.

A similar relation holds between the median and the mode: they lie within $3^{1/2} \approx 1.732$ standard deviations of each other:

$$\frac{|\tilde{X}-\text{mode}|}{\sigma} \leq 3^{1/2}.$$

Inequality Relating Means and Medians

If the distribution has finite variance, then the distance between the median and the mean is bounded by one standard deviation.

This bound was proved by Mallows, who used Jensen's inequality twice, as follows. We have,

$$\begin{aligned} |\mu - m| = |\mathrm{E}(X-m)| &\le \mathrm{E}(|X-m|) \\ &\le \mathrm{E}(|X-\mu|) \\ &\le \sqrt{\mathrm{E}\left((X-\mu)^2\right)} = \sigma. \end{aligned}$$

The first and third inequalities come from Jensen's inequality applied to the absolute-value function and the square function, which are each convex. The second inequality comes from the fact that a median minimizes the absolute deviation function,

$$a \mapsto \mathrm{E}(|X-a|).$$

This proof also follows directly from Cantelli's inequality. The result can be generalized to obtain a multivariate version of the inequality, as follows:

$$\begin{aligned} \|\mu - m\| = \|\mathrm{E}(X-m)\| \le \mathrm{E}\|X-m\| & \\ \le \mathrm{E}(\|X-\mu\|) & \\ \le \sqrt{\mathrm{E}\left(\|X-\mu\|^2\right)} = \sqrt{\operatorname{trace}\left(\operatorname{var}(X)\right)} & \end{aligned}$$

where m is a spatial median, that is, a minimizer of the function $a \mapsto \mathrm{E}(|X-a|)$. The spatial median is unique when the data-set's dimension is two or more. An alternative proof uses the one-sided Chebyshev inequality; it appears in an inequality on location and scale parameters.

Jensen's Inequality for Medians

Jensen's inequality states that for any random variable x with a finite expectation $E(x)$ and for any convex function f,

$$f[E(x)] \le E[f(x)]$$

It has been shown that if x is a real variable with a unique median m and f is a C function then,

$$f(m) \le \mathrm{Median}[f(x)]$$

A C function is a real valued function, defined on the set of real numbers R, with the property that for any real t,

$$f^{-1}\left((-\infty, t]\right) = \{x \in R \mid f(x) \le t\}$$

is a closed interval, a singleton or an empty set.

Medians for Samples

Efficient Computation of the Sample Median

Even though comparison-sorting n items requires $\Omega(n \log n)$ operations, selection algorithms can compute the k'th-smallest of n items with only $\Theta(n)$ operations. This includes the median, which is the (n/2)'th order statistic (or for an even number of samples, the arithmetic mean of the two middle order statistics).

Selection algorithms still have the downside of requiring $\Omega(n)$ memory, that is, they need to have the full sample (or a linear-sized portion of it) in memory. Because this, as well as the linear time requirement, can be prohibitive, several estimation procedures for the median have been developed. A simple one is the median of three rule, which estimates the median as the median of a three-element subsample; this is commonly used as a subroutine in the quicksort sorting algorithm, which uses an estimate of its input's median. A more robust estimator is Tukey's *ninther*, which is the median of three rule applied with limited recursion: if A is the sample laid out as an array,

$$\text{med3}(A) = \text{median}(A, A[n/2], A[n]),$$

then,

$$\text{ninther}(A) = \text{med3}(\text{med3}(A[1 \ldots 1/3n]), \text{med3}(A[1/3n \ldots 2/3n]), \text{med3}(A[2/3n \ldots n]))$$

The *remedian* is an estimator for the median that requires linear time but sub-linear memory, operating in a single pass over the sample.

Easy Explanation of the Sample Median

In individual series (if number of observation is very low) first one must arrange all the observations in order. Then count(n) is the total number of observation in given data.

If n is odd then Median (M) = value of ((n + 1)/2)th item term.

If n is even then Median (M) = value of [(n/2)th item term + (n/2 + 1)th item term]/2

For an Odd Number of Values

As an example, we will calculate the sample median for the following set of observations: 1, 5, 2, 8, 7.

Start by sorting the values: 1, 2, 5, 7, 8.

In this case, the median is 5 since it is the middle observation in the ordered list.

The median is the ((n + 1)/2)th item, where n is the number of values. For example, for the list {1, 2, 5, 7, 8}, we have n = 5, so the median is the ((5 + 1)/2)th item.

median = (6/2)th item

median = 3rd item

median = 5.

For an Even Number of Values

As an example, we will calculate the sample median for the following set of observations: 1, 6, 2, 8, 7, 2.

Start by sorting the values: 1, 2, 2, 6, 7, 8.

In this case, the arithmetic mean of the two middlemost terms is (2 + 6)/2 = 4. Therefore, the median is 4 since it is the arithmetic mean of the middle observations in the ordered list.

Sampling Distribution

The distributions of both the sample mean and the sample median were determined by Laplace. The distribution of the sample median from a population with a density function $f(x)$ is asymptotically normal with mean m and variance,

$$\frac{1}{4nf(m)^2}$$

where m is the median of $f(x)$ and n is the sample size. For normal samples, the density is $f(m) = 1/\sqrt{2\pi\sigma^2}$, thus for large samples the variance of the median equals $(\pi/2)\cdot(\sigma^2/n)$.

These results have also been extended. It is now known for the p-th quantile that the distribution of the sample p-th quantile is asymptotically normal around the p-th quantile with variance equal to:

$$\frac{p(1-p)}{nf(x_p)^2}$$

where $f(x_p)$ is the value of the distribution density at the p-th quantile.

Numerical Experimentation

In the case of a discrete variable, the sampling distribution of the median for small-samples can be investigated as follows. We take the sample size to be an odd number

$N = 2n+1$. If a given value v is to be the median of the sample then two conditions must be satisfied. The first is that at most observations can have a value of $v-1$ or less. The second is that at most n observations can have a value of $v+1$ or more. Let i be the number of observations that have a value of $v+1$ or less and let $v-1$ be the number of observations that have a value of $v+1$ or more. Then i and k both have a minimum value of 0 and a maximum of n. If an observation has a value below v, it is not relevant how far below v it is and conversely, if an observation has a value above v, it is not relevant how far above v it is. We can therefore represent the observations as following a trinomial distribution with probabilities $F(v-1)$, $f(v)$ and $1-F(v)$. The probability that the median m will have a value v is then given by,

$$\Pr(m = v) = \sum_{i=0}^{n}\sum_{k=0}^{n} \frac{N!}{i!(N-i-k)!k!}[F(v-1)]^{i}[f(v)]^{N-i-k}[1-F(v)]^{k}.$$

Summing this over all values of v defines a proper distribution and gives a unit sum. In practice, the function $f(v)$ will often not be known but it can be estimated from an observed frequency distribution. An example is given in the following table where the actual distribution is not known but a sample of 3,800 observations allows a sufficiently accurate assessment of $f(v)$.

v	0	0.5	1	1.5	2	2.5	3	3.5	4	4.5	5
f(v)	0.000	0.008	0.010	0.013	0.083	0.108	0.328	0.220	0.202	0.023	0.005
F(v)	0.000	0.008	0.018	0.031	0.114	0.222	0.550	0.770	0.972	0.995	1.000

Using these data it is possible to investigate the effect of sample size on the standard errors of the mean and median. The observed mean is 3.16, the observed raw median is 3 and the observed interpolated median is 3.174. The following table gives some comparison statistics. The standard error of the median is given both from the above expression for $pr(m = v)$ and from the asymptotic approximation given earlier.

	Sample Size			
Statistic	3	9	15	21
Expected value of median	3.198	3.191	3.174	3.161
Standard error of median (above formula)	0.482	0.305	0.257	0.239
Standard error of median (asymptotic approximation)	0.879	0.508	0.393	0.332
Standard error of mean	0.421	0.243	0.188	0.159

The expected value of the median falls slightly as sample size increases while, as would be expected, the standard errors of both the median and the mean are proportionate to the inverse square root of the sample size. The asymptotic approximation errs on the side of caution by overestimating the standard error.

In the case of a continuous variable, the following argument can be used. If a given value v is to be the median, then one observation must take the value v. The elemental probability of this is $f(v)dv$. Then, of the remaining $2n$ observations, exactly n of them must be above v and the remaining n below. The probability of this is the n th term of a binomial distribution with parameters $F(v)$ and $2n$. Finally we multiply by $2n+1$ since any of the observations in the sample can be the median observation. Hence the elemental probability of the median at the point v is given by,

$$f(v)\frac{(2n)!}{n!n!}[F(v)]^n[1-F(v)]^n(2n+1)dv.$$

Now we introduce the beta function. For integer arguments α and β, this can be expressed as $\mathrm{B}(\alpha,\beta)=(\alpha-1)!(\beta-1)!/(\alpha+\beta-1)!$. Also, $f(v)=dF(v)/dv$. Using these relationships and setting both α and β, equal to $(n+1)$ allows the last expression to be written as,

$$\frac{[F(v)]^n[1-F(v)]^n}{\mathrm{B}(n+1,n+1)}dF(v)$$

Hence the density function of the median is a symmetric beta distribution over the unit interval which supports $F(v)$. Its mean, as we would expect, is 0.5 and its variance is $1/(4(N+2))$. The corresponding variance of the sample median is,

$$\frac{1}{4(N+2)f(m)^2}.$$

However this finding can only be used if the density function $f(v)$ is known or can be assumed. As this will not always be the case, the median variance has to be estimated sometimes from the sample data.

Estimation of Variance from Sample Data

The value of $(2f(x))^{-2}$ —the asymptotic value of $n^{-\frac{1}{2}}(v-m)$ where v is the population median—has been studied by several authors. The standard "delete one" jackknife method produces inconsistent results. An alternative—the "delete k" method—where k grows with the sample size has been shown to be asymptotically consistent. This method may be computationally expensive for large data sets. A bootstrap estimate is known to be consistent, but converges very slowly (order of $n^{-\frac{1}{4}}$). Other methods have been proposed but their behavior may differ between large and small samples.

Efficiency

The efficiency of the sample median, measured as the ratio of the variance of the mean to the variance of the median, depends on the sample size and on the underlying population distribution. For a sample of size $N = 2n+1$ from the normal distribution, the efficiency for large N is,

$$\frac{2}{\pi}\frac{N+2}{N}$$

The efficiency tends to $\frac{2}{\pi}$ as N tends to infinity.

In other words, the relative variance of the median will be $\pi/2 \approx 1.57,$, or 57% greater than the variance of the mean – the standard error of the median will be 25% greater than that of the mean.

Other Estimators

For univariate distributions that are *symmetric* about one median, the Hodges–Lehmann estimator is a robust and highly efficient estimator of the population median.

If data are represented by a statistical model specifying a particular family of probability distributions, then estimates of the median can be obtained by fitting that family of probability distributions to the data and calculating the theoretical median of the fitted distribution. Pareto interpolation is an application of this when the population is assumed to have a Pareto distribution.

Coefficient of Dispersion

The coefficient of dispersion (CD) is defined as the ratio of the average absolute deviation from the median to the median of the data. It is a statistical measure used by the states of Iowa, New York and South Dakota in estimating dues taxes. In symbols,

$$CD = \frac{1}{n}\frac{\sum |m-x|}{m}$$

where *n* is the sample size, *m* is the sample median and *x* is a variate. The sum is taken over the whole sample.

Confidence intervals for a two-sample test in which the sample sizes are large have been derived by Bonett and Seier. This test assumes that both samples have the same median but differ in the dispersion around it. The confidence interval (CI) is bounded inferiorly by,

$$\exp\left[\log\left(\frac{t_a}{t_b}\right) - z_\alpha\left(\mathrm{var}\left[\log\left(\frac{t_a}{t_b}\right)\right]\right)^{1/2}\right]$$

where t_j is the mean absolute deviation of the jth sample, var() is the variance and z_α is the value from the normal distribution for the chosen value of α: for $\alpha = 0.05$, $z_\alpha = 1.96$. The following formulae are used in the derivation of these confidence intervals,

$$\text{var}[\log(t_a)] = \frac{1}{n}\left[\frac{s_a^2}{t_a^2} + \left(\frac{x_a - \overline{x}}{t_a}\right)^2 - 1\right]$$

$$\text{var}\left[\log\left(\frac{t_a}{t_b}\right)\right] = \text{var}[\log(t_a)] + \text{var}[\log(t_b)] - 2r(\text{var}[\log(t_a)]\,\text{var}[\log(t_b)])^{1/2}$$

where r is the Pearson correlation coefficient between the squared deviation scores,

$$d_{ia} = | x_{ia} - \overline{x}_a | \text{ and } d_{ib} = | x_{ib} - \overline{x}_b |$$

a and b here are constants equal to 1 and 2, x is a variate and s is the standard deviation of the sample.

Multivariate Median

When the dimension is two or higher, there are multiple concepts that extend the definition of the univariate median; each such multivariate median agrees with the univariate median when the dimension is exactly one.

Marginal Median

The marginal median is defined for vectors defined with respect to a fixed set of coordinates. A marginal median is defined to be the vector whose components are univariate medians. The marginal median is easy to compute, and its properties were studied by Puri and Sen.

Centerpoint

An alternative generalization of the median in higher dimensions is the centerpoint.

Other Median-related Concepts

Interpolated Median

When dealing with a discrete variable, it is sometimes useful to regard the observed values as being midpoints of underlying continuous intervals. An example of this is a Likert scale, on which opinions or preferences are expressed on a scale with a set number of possible responses. If the scale consists of the positive integers, an observation of 3 might be regarded as representing the interval from 2.50 to 3.50. It is possible to estimate the median of the underlying variable. If, say, 22% of the observations are of value 2 or below and 55.0% are of 3 or below (so 33% have the value 3), then the median m is 3 since the median is the smallest value of x for which $F(x)$ is greater than a half. But the interpolated median is somewhere between 2.50 and 3.50. First we add half of the interval width w to the median to get the upper bound of the median interval. Then we subtract that proportion of the interval width which equals the proportion of the 33% which lies above the 50% mark. In other words, we split up the interval width pro rata to the numbers of observations. In this case, the 33% is split into 28% below the median and 5% above it so we subtract 5/33 of the interval width from the upper bound of 3.50 to give an interpolated median of 3.35. More formally, if the values $f(x)$ are known, the interpolated median can be calculated from,

$$m_{\text{int}} = m + w\left[\frac{1}{2} - \frac{F(m) - \frac{1}{2}}{f(m)}\right].$$

Alternatively, if in an observed sample there are k scores above the median category, j scores in it and i scores below it then the interpolated median is given by,

$$m_{\text{int}} = m - \frac{w}{2}\left[\frac{k-i}{j}\right].$$

Pseudo-median

For univariate distributions that are *symmetric* about one median, the Hodges–Lehmann estimator is a robust and highly efficient estimator of the population median; for non-symmetric distributions, the Hodges–Lehmann estimator is a robust and highly efficient estimator of the population *pseudo-median*, which is the median of a symmetrized distribution and which is close to the population median. The Hodges–Lehmann estimator has been generalized to multivariate distributions.

Variants of Regression

The Theil–Sen estimator is a method for robust linear regression based on finding medians of slopes.

Median Filter

In the context of image processing of monochrome raster images there is a type of noise, known as the salt and pepper noise, when each pixel independently becomes black (with some small probability) or white (with some small probability), and is unchanged otherwise (with the probability close to 1). An image constructed of median values of neighborhoods (like 3×3 square) can effectively reduce noise in this case.

Cluster Analysis

In cluster analysis, the k-medians clustering algorithm provides a way of defining clusters, in which the criterion of maximising the distance between cluster-means that is used in k-means clustering, is replaced by maximising the distance between cluster-medians.

Median–median Line

This is a method of robust regression. The idea dates back to Wald in 1940 who suggested dividing a set of bivariate data into two halves depending on the value of the independent parameter x: A left half with values less than the median and a right half with values greater than the median. He suggested taking the means of the dependent x and independent variables of the left and the right halves and estimating the slope of the line joining these two points. The line could then be adjusted to fit the majority of the points in the data set.

Nair and Shrivastava in 1942 suggested a similar idea but instead advocated dividing the sample into three equal parts before calculating the means of the subsamples. Brown and Mood in 1951 proposed the idea of using the medians of two subsamples rather the means. Tukey combined these ideas and recommended dividing the sample into three equal size subsamples and estimating the line based on the medians of the subsamples.

Median-unbiased Estimators

Any *mean*-unbiased estimator minimizes the risk (expected loss) with respect to the squared-error loss function, as observed by Gauss. A *median*-unbiased estimator minimizes the risk with respect to the absolute-deviation loss function, as observed by Laplace. Other loss functions are used in statistical theory, particularly in robust statistics.

The theory of median-unbiased estimators was revived by George W. Brown in 1947:

> "An estimate of a one-dimensional parameter θ will be said to be median-unbiased if, for fixed θ, the median of the distribution of the estimate is at the value θ; i.e., the estimate underestimates just as often as it overestimates. This requirement seems for most purposes to accomplish as much as the mean-unbiased requirement and has the additional property that it is invariant under one-to-one transformation".

Further properties of median-unbiased estimators have been reported. Median-unbiased estimators are invariant under one-to-one transformations.

There are methods of constructing median-unbiased estimators that are optimal (in a sense analogous to the minimum-variance property for mean-unbiased estimators). Such constructions exist for probability distributions having monotone likelihood-functions.

One such procedure is an analogue of the Rao–Blackwell procedure for mean-unbiased estimators: The procedure holds for a smaller class of probability distributions than does the Rao—Blackwell procedure but for a larger class of loss functions.

MODE

The mode of a set of data values is the value that appears most often. If X is a discrete random variable, the mode is the value x (i.e, X = x) at which the probability mass function takes its maximum value. In other words, it is the value that is most likely to be sampled.

Like the statistical mean and median, the mode is a way of expressing, in a (usually) single number, important information about a random variable or a population. The numerical value of the mode is the same as that of the mean and median in a normal distribution, and it may be very different in highly skewed distributions.

The mode is not necessarily unique to a given discrete distribution, since the probability mass function may take the same maximum value at several points x_1, x_2, etc. The most extreme case occurs in uniform distributions, where all values occur equally frequently.

When the probability density function of a continuous distribution has multiple local maxima it is common to refer to all of the local maxima as modes of the distribution. Such a continuous distribution is called multimodal (as opposed to unimodal). A mode of a continuous probability distribution is often considered to be any value x at which its probability density function has a locally maximum value, so any peak is a mode.

In symmetric unimodal distributions, such as the normal distribution, the mean (if defined), median and mode all coincide. For samples, if it is known that they are drawn from a symmetric unimodal distribution, the sample mean can be used as an estimate of the population mode.

Mode of a Sample

The mode of a sample is the element that occurs most often in the collection. For example, the mode of the sample [1, 3, 6, 6, 6, 6, 7, 7, 12, 12, 17] is 6. Given the list of data [1, 1, 2, 4, 4] the mode is not unique – the dataset may be said to be bimodal, while a set with more than two modes may be described as multimodal.

For a sample from a continuous distribution, such as [0.935..., 1.211..., 2.430..., 3.668..., 3.874...], the concept is unusable in its raw form, since no two values will be exactly the same, so each value will occur precisely once. In order to estimate the mode of the underlying distribution, the usual practice is to discretize the data by assigning frequency values to intervals of equal distance, as for making a histogram, effectively replacing the values by the midpoints of the intervals they are assigned to. The mode is then the value where the histogram reaches its peak. For small or middle-sized samples the outcome of this procedure is sensitive to the choice of interval width if chosen too narrow or too wide; typically one should have a sizable fraction of the data concentrated in a relatively small number of intervals (5 to 10), while the fraction of the data falling outside these intervals is also sizable. An alternate approach is kernel density estimation, which essentially blurs point samples to produce a continuous estimate of the probability density function which can provide an estimate of the mode.

The following MATLAB (or Octave) code example computes the mode of a sample:

```
X = sort(x);

indices = find(diff([X; realmax]) > 0); % indices where repeated values
change

[modeL,i] = max (diff([0; indices])); % longest persistence length of
repeated values

mode = X(indices(i));
```

The algorithm requires as a first step to sort the sample in ascending order. It then computes the discrete derivative of the sorted list, and finds the indices where this derivative is positive. Next it computes the discrete derivative of this set of indices, locating the maximum of this derivative of indices, and finally evaluates the sorted sample at the point where that maximum occurs, which corresponds to the last member of the stretch of repeated values.

Comparison of Mean, Median and Mode

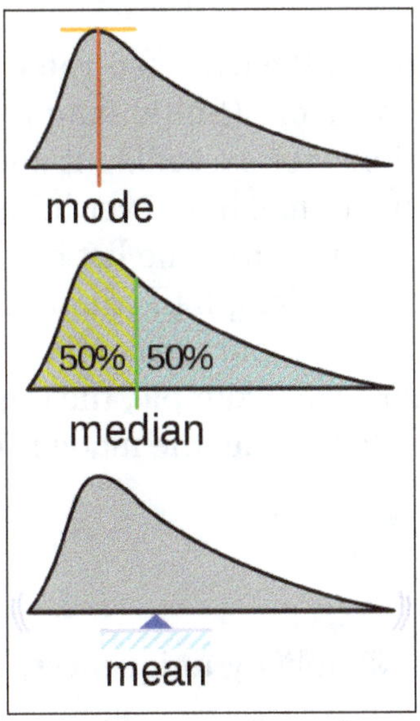

Geometric visualisation of the mode, median and mean of an arbitrary probability density function.

Comparison of common averages of values { 1, 2, 2, 3, 4, 7, 9 }			
Type	Description	Example	Result
Arithmetic mean	Sum of values of a data set divided by number of values:	(1+2+2+3+4+7+9) / 7	4
Median	Middle value separating the greater and lesser halves of a data set	1, 2, 2, 3, 4, 7, 9	3
Mode	Most frequent value in a data set	1, 2, 2, 3, 4, 7, 9	2

Use

Unlike mean and median, the concept of mode also makes sense for "nominal data" (i.e., not consisting of numerical values in the case of mean, or even of ordered values in the case of median). For example, taking a sample of Korean family names, one might find that "Kim" occurs more often than any other name. Then "Kim" would be the mode of the sample. In any voting system where a plurality determines victory, a single modal value determines the victor, while a multi-modal outcome would require some tie-breaking procedure to take place.

Unlike median, the concept of mode makes sense for any random variable assuming values from a vector space, including the real numbers (a one-dimensional vector space) and the integers (which can be considered embedded in the reals). For example, a distribution of points in the plane will typically have a mean and a mode, but the concept of median does not apply. The median makes sense when there is a linear order on the possible values. Generalizations of the concept of median to higher-dimensional spaces are the geometric median and the centerpoint.

Uniqueness and Definedness

For some probability distributions, the expected value may be infinite or undefined, but if defined, it is unique. The mean of a (finite) sample is always defined. The median is the value such that the fractions not exceeding it and not falling below it are each at least 1/2. It is not necessarily unique, but never infinite or totally undefined. For a data sample it is the "halfway" value when the list of values is ordered in increasing value, where usually for a list of even length the numerical average is taken of the two values closest to "halfway". Finally, as said before, the mode is not necessarily unique. Certain pathological distributions (for example, the Cantor distribution) have no defined mode at all. For a finite data sample, the mode is one (or more) of the values in the sample.

Properties

Assuming definedness, and for simplicity uniqueness, the following are some of the most interesting properties.

- All three measures have the following property: If the random variable (or each value from the sample) is subjected to the linear or affine transformation, which replaces X by $aX+b$, so are the mean, median and mode.

- Except for extremely small samples, the mode is insensitive to "outliers" (such as occasional, rare, false experimental readings). The median is also very robust in the presence of outliers, while the mean is rather sensitive.
- In continuous unimodal distributions the median often lies between the mean and the mode, about one third of the way going from mean to mode. In a formula, median ≈ (2 × mean + mode)/3. This rule, due to Karl Pearson, often applies to slightly non-symmetric distributions that resemble a normal distribution, but it is not always true and in general the three statistics can appear in any order.
- For unimodal distributions, the mode is within standard deviations of the mean, and the root mean square deviation about the mode is between the standard deviation and twice the standard deviation.

Example for a Skewed Distribution

An example of a skewed distribution is personal wealth: Few people are very rich, but among those some are extremely rich. However, many are rather poor.

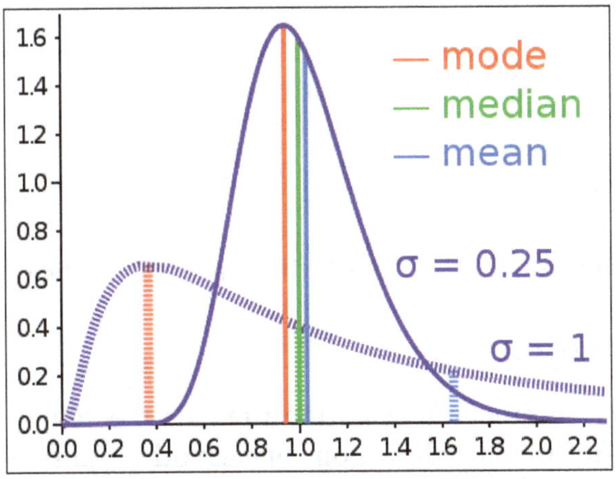

Comparison of mean, median and mode of two log-normal distributions with different skewness.

A well-known class of distributions that can be arbitrarily skewed is given by the log-normal distribution. It is obtained by transforming a random variable X having a normal distribution into random variable $Y = e^X$. Then the logarithm of random variable Y is normally distributed, hence the name.

Taking the mean μ of X to be 0, the median of Y will be 1, independent of the standard deviation σ of X. This is so because X has a symmetric distribution, so its median is also 0. The transformation from X to Y is monotonic, and so we find the median $e^0 = 1$ for Y.

When X has standard deviation σ = 0.25, the distribution of Y is weakly skewed. Using formulas for the log-normal distribution, we find:

$$\begin{array}{llll} \text{mean} & = e^{\mu+\sigma^2/2} & = e^{0+0.25^2/2} & \approx 1.032 \\ \text{mode} & = e^{\mu-\sigma^2} & = e^{0-0.25^2} & \approx 0.939 \\ \text{median} & = e^{\mu} & = e^{0} & = 1 \end{array}$$

Indeed, the median is about one third on the way from mean to mode.

When X has a larger standard deviation, σ = 1, the distribution of Y is strongly skewed. Now

$$\begin{array}{llll} \text{mean} & = e^{\mu+\sigma^2/2} & = e^{0+1^2/2} & \approx 1.649 \\ \text{mode} & = e^{\mu-\sigma^2} & = e^{0-1^2} & \approx 0.368 \\ \text{median} & = e^{\mu} & = e^{0} & = 1 \end{array}$$

Here, Pearson's rule of thumb fails.

Van Zwet Condition

Van Zwet derived an inequality which provides sufficient conditions for this inequality to hold. The inequality

Mode ≤ Median ≤ Mean

holds if

F(Median - x) + F(Median + x) ≥ 1

for all x where F() is the cumulative distribution function of the distribution.

Unimodal Distributions

It can be shown for a unimodal distribution that the median $\tilde{X}$ and the mean $\bar{X}$ lie within $(3/5)^{1/2} \approx 0.7746$ standard deviations of each other. In symbols,

$$\frac{|\tilde{X} - \bar{X}|}{\sigma} \leq (3/5)^{1/2}$$

where $|\cdot|$ is the absolute value.

A similar relation holds between the median and the mode: they lie within $3^{1/2} \approx 1.732$ standard deviations of each other:

$$\frac{|\tilde{X} - \text{mode}|}{\sigma} \leq 3^{1/2}.$$

RANGE

In statistics, the range of a set of data is the difference between the largest and smallest values. Difference here is specific, the range of a set of data is the result of subtracting the smallest value from largest value.

However, in descriptive statistics, this concept of range has a more complex meaning. The range is the size of the smallest interval (statistics) which contains all the data and provides an indication of statistical dispersion. It is measured in the same units as the data. Since it only depends on two of the observations, it is most useful in representing the dispersion of small data sets.

For Continuous IID Random Variables

For n independent and identically distributed continuous random variables $X_1, X_2, \ldots, X_n$ with cumulative distribution function G(x) and probability density function g(x). Let T denote the range of a sample of size n from a population with distribution function $G(x)$.

Distribution

The range has cumulative distribution function,

$$F(t) = n\int_{-\infty}^{\infty} g(x)[G(x+t) - G(x)]^{n-1}\mathrm{d}x.$$

Gumbel notes that the "beauty of this formula is completely marred by the facts that, in general, we cannot express $G(x + t)$ by $G(x)$, and that the numerical integration is lengthy and tiresome."

If the distribution of each X_i is limited to the right (or left) then the asymptotic distribution of the range is equal to the asymptotic distribution of the largest (smallest) value. For more general distributions the asymptotic distribution can be expressed as a Bessel function.

Moments

The mean range is given by,

$$n\int_0^1 x(G)[G^{n-1} - (1-G)^{n-1}]\mathrm{d}G$$

where $x(G)$ is the inverse function. In the case where each of the X_i has a standard normal distribution, the mean range is given by,

$$\int_{-\infty}^{\infty} (1 - (1-\Phi(x))^n - \Phi(x)^n)\mathrm{d}x.$$

For Continuous Non-IID Random Variables

For n nonidentically distributed independent continuous random variables X_1, X_2, ..., X_n with cumulative distribution functions $G_1(x)$, $G_2(x)$, ..., $G_n(x)$ and probability density functions $g_1(x)$, $g_2(x)$, ..., $g_n(x)$, the range has cumulative distribution function,

$$F(t)=\sum_{i=1}^{n}\int_{-\infty}^{\infty} g_i(x) \prod_{j=1, j\neq i}^{n} [G_j(x+t)-G_j(x)]\mathrm{d}x.$$

For Discrete IID Random Variables

For n independent and identically distributed discrete random variables X_1, X_2, ..., X_n with cumulative distribution function $G(x)$ and probability mass function $g(x)$ the range of the X_i is the range of a sample of size n from a population with distribution function $G(x)$. We can assume without loss of generality that the support of each X_i is $\{1,2,3,...,N\}$ where N is a positive integer or infinity.

Distribution

The range has probability mass function,

$$f(t)=\begin{cases} \sum_{x-1}^{N}\left[g(x)\right]^n & t=0 \\ \sum_{x=1}^{N-t}\begin{pmatrix} \left[G(x+t)-G(x-1)\right]^n \\ -\left[G(x+t)-G(x)\right]^n \\ -\left[G(x+t-1)-G(x-1)\right]^n \\ +\left[G(x+t-1)-G(x)\right]^n \end{pmatrix} & t=1,2,3...,N-1 \end{cases}$$

Example:

If we suppose that $g(x) = 1/N$, the discrete uniform distribution for all x, then we find,

$$f(t)=\begin{cases} \dfrac{1}{N^{n-1}} & t=0 \\ \sum_{x=1}^{N-t}\left([\frac{t+1}{N}]^n-2[\frac{t}{N}]^n+[\frac{t-1}{N}]^n\right) & t=1,2,3\ldots,N-1. \end{cases}$$

Derivation

The probability of having a specific range value, t, can be determined by adding the probabilities of having two samples differing by t, and every other sample having a value between the two extremes. The probability of one sample having a value of x is $n * g(x)$. The probability of another having a value t greater than x is:

$$(n-1)g(x+t).$$

The probability of all other values lying between these two extremes is:

$$\left(\int_{x}^{x+1} g(x)\mathrm{d}x\right)^{n-2} = \left(G(x+t)-G(x)\right)^{n-2}.$$

Combining the three together yields:

$$f(t) = n(n-1)\int_{-\infty}^{\infty} g(x)g(x+t)\left[G(x+t)-G(x)\right]^{n-2} dx$$

Related Quantities

The range is a simple function of the sample maximum and minimum and these are specific examples of order statistics. In particular, the range is a linear function of order statistics, which brings it into the scope of L-estimation.

Interquartile Range

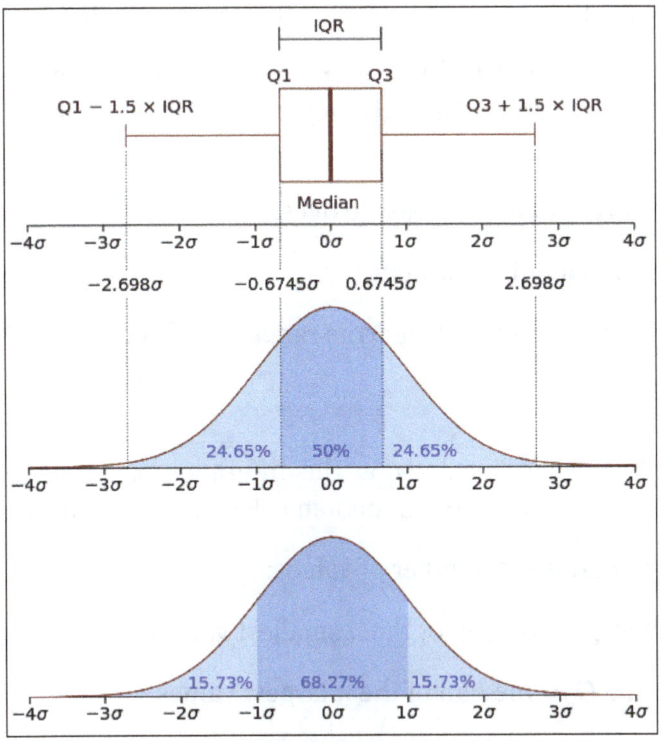

Boxplot (with an interquartile range) and a probability density function (pdf) of a Normal N(0,σ²) Population.

In descriptive statistics, the interquartile range (IQR), also called the midspread or middle 50%, or technically H-spread, is a measure of statistical dispersion, being equal to the difference between 75th and 25th percentiles, or between upper and lower quartiles, IQR = $Q_3 - Q_1$.

In other words, the IQR is the first quartile subtracted from the third quartile; these quartiles can be clearly seen on a box plot on the data. It is a trimmed estimator, defined as the 25% trimmed range, and is a commonly used robust measure of scale.

The IQR is a measure of variability, based on dividing a data set into quartiles. Quartiles divide a rank-ordered data set into four equal parts. The values that separate parts are called the first, second, and third quartiles; and they are denoted by Q1, Q2, and Q3, respectively.

Use

Unlike total range, the interquartile range has a breakdown point of 25%, and is thus often preferred to the total range.

The IQR is used to build box plots, simple graphical representations of a probability distribution.

The IQR is used in businesses as a marker for their income rates.

For a symmetric distribution (where the median equals the midhinge, the average of the first and third quartiles), half the IQR equals the median absolute deviation (MAD).

The median is the corresponding measure of central tendency.

The IQR can be used to identify outliers.

The quartile deviation or semi-interquartile range is defined as half the IQR.

Algorithm

The IQR of a set of values is calculated as the difference between the upper and lower quartiles, Q_3 and Q_1. Each quartile is a median calculated as follows.

Given an even *2n* or odd *2n+1* number of values:

First quartile Q_1 = median of the *n* smallest values,

Third quartile Q_3 = median of the *n* largest values.

The *second quartile* Q_2 is the same as the ordinary median.

Data Set in a Table

The following table has 13 rows, and follows the rules for the odd number of entries.

i	x[i]	Median	Quartile
1	7	Q_2=87 (median of whole table)	Q_1=31 (median of upper half, from row 1 to 6)
2	7		
3	31		
4	31		
5	47		
6	75		
7	87		
8	115		Q_3=119 (median of lower half, from row 8 to 13)
9	116		
10	119		
11	119		
12	155		
13	177		

For the data in this table the interquartile range is IQR = $Q_3 - Q_1$ = 119 - 31 = 88.

Data Set in a Plain-text Box Plot

```
                               +-----+-+
                 * |-----------| | |-----------|
                               +-----+-+


+---+---+---+---+---+---+---+---+---+---+---+---+ number line
0   1   2   3   4   5   6   7   8   9  10   11  12
```

For the data set in this box plot:

- lower (first) quartile Q_1 = 7
- median (second quartile) Q_2 = 8.5
- upper (third) quartile Q_3 = 9
- interquartile range, IQR = Q_3 - Q_1 = 2
- lower 1.5*IQR whisker = Q_1 - 1.5 * IQR = 7 - 3 = 4. (If there is no data point at 4, then the lowest point greater than 4.)
- upper 1.5*IQR whisker = Q_3 + 1.5 * IQR = 9 + 3 = 12. (If there is no data point at 12, then the highest point less than 12.)

This means the 1.5*IQR whiskers can be uneven in lengths.

Distributions

The interquartile range of a continuous distribution can be calculated by integrating the probability density function (which yields the cumulative distribution function—any other means of calculating the CDF will also work). The lower quartile, Q_1, is a number such that integral of the PDF from -∞ to Q_1 equals 0.25, while the upper quartile, Q_3, is such a number that the integral from -∞ to Q_3 equals 0.75; in terms of the CDF, the quartiles can be defined as follows:

$$Q_1 = \text{CDF}^{-1}(0.25),$$

$$Q_3 = \text{CDF}^{-1}(0.75),$$

where CDF^{-1} is the quantile function.

The interquartile range and median of some common distributions are shown below:

Distribution	Median	IQR
Normal	μ	2 Φ−1(0.75)σ ≈ 1.349σ ≈ (27/20)σ
Laplace	μ	2b ln(2) ≈ 1.386b
Cauchy	μ	2γ

Interquartile Range Test for Normality of Distribution

The IQR, mean, and standard deviation of a population *P* can be used in a simple test of whether or not *P* is normally distributed, or Gaussian. If *P* is normally distributed, then the standard score of the first quartile, z_1, is −0.67, and the standard score of the third quartile, z_3, is +0.67. Given *mean* = *X* and *standard deviation* = σ for *P*, if *P* is normally distributed, the first quartile,

$$Q_1 = (\sigma z_1) + X$$

and the third quartile,

$$Q_3 = (\sigma z_3) + X$$

If the actual values of the first or third quartiles differ substantially from the calculated values, *P* is not normally distributed. However, a normal distribution can be trivially perturbed to maintain its Q1 and Q2 std. scores at 0.67 and −0.67 and not be normally distributed (so the above test would produce a false positive). A better test of normality, such as Q-Q plot would be indicated here.

Outliers

The interquartile range is often used to find outliers in data. Outliers here are defined as observations that fall below Q1 − 1.5 IQR or above Q3 + 1.5 IQR. In a boxplot, the highest and lowest occurring value within this limit are indicated by *whiskers* of the box (frequently with an additional bar at the end of the whisker) and any outliers as individual points.

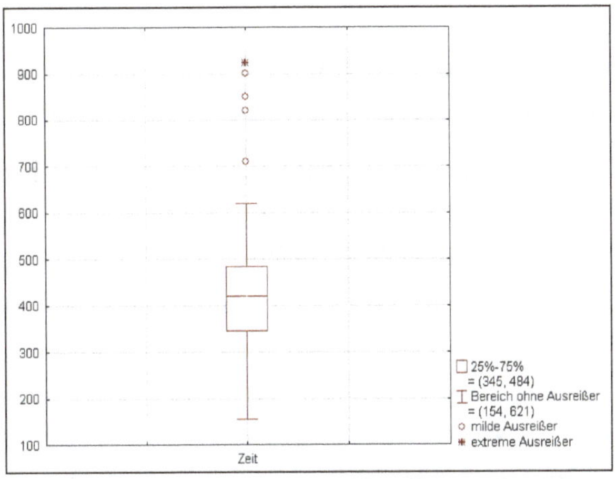

Box-and-whisker plot with four mild outliers and one extreme outlier. In this chart, outliers are defined as mild above Q3 + 1.5 IQR and extreme above Q3 + 3 IQR.

STATISTICAL DISPERSION

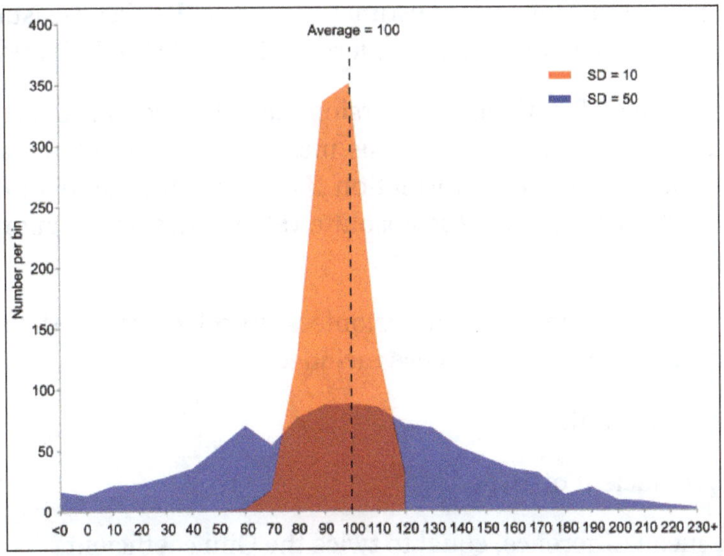

Example of samples from two populations with the same mean but different dispersion. The blue population is much more dispersed than the red population.

In statistics, dispersion (also called variability, scatter, or spread) is the extent to which a distribution is stretched or squeezed. Common examples of measures of statistical dispersion are the variance, standard deviation, and interquartile range.

Dispersion is contrasted with location or central tendency, and together they are the most used properties of distributions.

Measures

A measure of statistical dispersion is a nonnegative real number that is zero if all the data are the same and increases as the data become more diverse.

Most measures of dispersion have the same units as the quantity being measured. In other words, if the measurements are in metres or seconds, so is the measure of dispersion. Examples of dispersion measures include:

- Standard deviation.
- Interquartile range (IQR).
- Range.
- Mean absolute difference (also known as Gini mean absolute difference).
- Median absolute deviation (MAD).
- Average absolute deviation (or simply called average deviation).
- Distance standard deviation.

These are frequently used (together with scale factors) as estimators of scale parameters, in which capacity they are called estimates of scale. Robust measures of scale are those unaffected by a small number of outliers, and include the IQR and MAD.

All the above measures of statistical dispersion have the useful property that they are *location-invariant* and *linear in scale*. This means that if a random variable X has a dispersion of S_X then a linear transformation $Y = aX + b$ for real a and b should have dispersion $S_Y = |a|S_X$, where $|a|$ is the absolute value of a, that is, ignores a preceding negative sign $-$.

Other measures of dispersion are dimensionless. In other words, they have no units even if the variable itself has units. These include:

- Coefficient of variation.
- Quartile coefficient of dispersion.
- Relative mean difference, equal to twice the Gini coefficient.
- Entropy: While the entropy of a discrete variable is location-invariant and scale-independent, and therefore not a measure of dispersion in the above sense, the entropy of a continuous variable is location invariant and additive in scale: If Hz is the entropy of continuous variable z and $y=ax+b$, then $Hy=Hx+log(a)$.

There are other measures of dispersion:

- Variance (the square of the standard deviation) – location-invariant but not linear in scale.
- Variance-to-mean ratio – mostly used for count data when the term coefficient of dispersion is used and when this ratio is dimensionless, as count data are themselves dimensionless, not otherwise.

Some measures of dispersion have specialized purposes, among them the Allan variance and the Hadamard variance.

For categorical variables, it is less common to measure dispersion by a single number. One measure that does so is the discrete entropy.

Sources

In the physical sciences, such variability may result from random measurement errors: instrument measurements are often not perfectly precise, i.e., reproducible, and there is additional inter-rater variability in interpreting and reporting the measured results. One may assume that the quantity being measured is stable, and that the variation between measurements is due to observational error. A system of a large number of particles is characterized by the mean values of a relatively few number of macroscopic quantities such as temperature, energy, and density. The standard deviation is an important measure in fluctuation theory, which explains many physical phenomena, including why the sky is blue.

In the biological sciences, the quantity being measured is seldom unchanging and stable, and the variation observed might additionally be *intrinsic* to the phenomenon: It may be due to *inter-individual variability*, that is, distinct members of a population differing from each other. Also, it may be due to *intra-individual variability*, that is, one and the same subject differing in tests taken at different times or in other differing conditions. Such types of variability are also seen in the arena of manufactured products; even there, the meticulous scientist finds variation.

In economics, finance, and other disciplines, regression analysis attempts to explain the dispersion of a dependent variable, generally measured by its variance, using one or more independent variables each of which itself has positive dispersion. The fraction of variance explained is called the coefficient of determination.

A Partial Ordering of Dispersion

A mean-preserving spread (MPS) is a change from one probability distribution A to another probability distribution B, where B is formed by spreading out one or more portions of A's probability density function while leaving the mean (the expected value) unchanged. The concept of a mean-preserving spread provides a partial ordering of probability distributions according to their dispersions: of two probability distributions, one may be ranked as having more dispersion than the other, or alternatively neither may be ranked as having more dispersion.

SKEWNESS

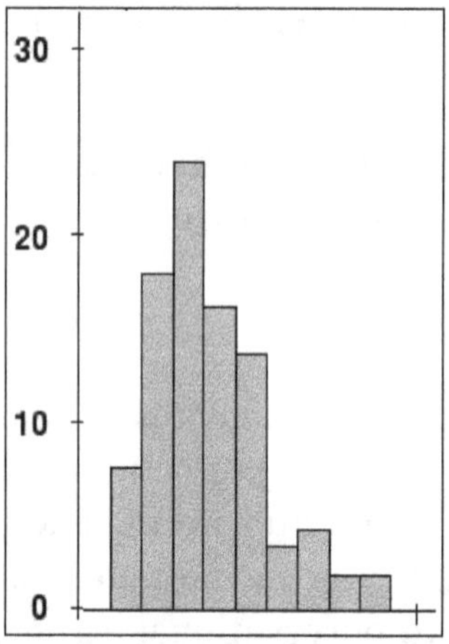

Example distribution with non-zero (positive) skewness.
These data are from experiments on wheat grass growth.

In probability theory and statistics, skewness is a measure of the asymmetry of the probability distribution of a real-valued random variable about its mean. The skewness value can be positive or negative, or undefined.

For a unimodal distribution, negative skew commonly indicates that the *tail* is on the left side of the distribution, and positive skew indicates that the tail is on the right. In cases where one tail is long but the other tail is fat, skewness does not obey a simple rule. For example, a zero value means that the tails on both sides of the mean balance out overall; this is the case for a symmetric distribution, but can also be true for an asymmetric distribution where one tail is long and thin, and the other is short but fat.

Consider the two distributions in the figure just below. Within each graph, the values on the right side of the distribution taper differently from the values on the left side. These tapering sides are called *tails*, and they provide a visual means to determine which of the two kinds of skewness a distribution has:

- *Negative skew*: The left tail is longer; the mass of the distribution is concentrated on the right of the figure. The distribution is said to be *left-skewed, left-tailed,* or *skewed to the left*, despite the fact that the curve itself appears to be skewed or leaning to the right; *left* instead refers to the left tail being drawn out and, often, the mean being skewed to the left of a typical center of the data. A left-skewed distribution usually appears as a *right-leaning* curve.

- *Positive skew*: The right tail is longer; the mass of the distribution is concentrated on the left of the figure. The distribution is said to be *right-skewed*, *right-tailed*, or *skewed to the right*, despite the fact that the curve itself appears to be skewed or leaning to the left; *right* instead refers to the right tail being drawn out and, often, the mean being skewed to the right of a typical center of the data. A right-skewed distribution usually appears as a *left-leaning* curve.

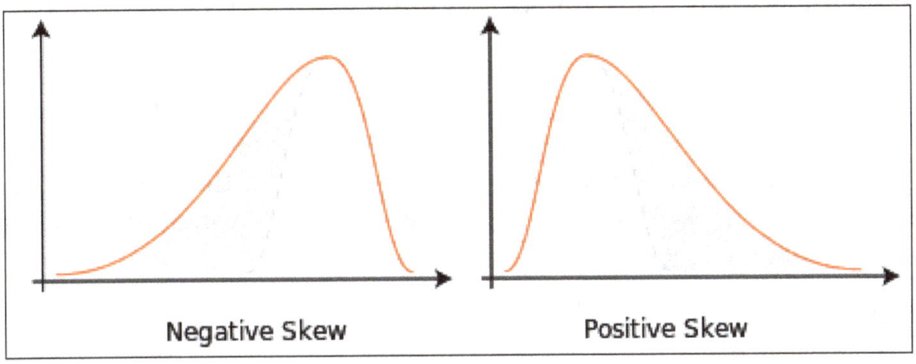

Skewness in a data series may sometimes be observed not only graphically but by simple inspection of the values. For instance, consider the numeric sequence (49, 50, 51), whose values are evenly distributed around a central value of 50. We can transform this sequence into a negatively skewed distribution by adding a value far below the mean, which is probably a negative outlier, e.g. (40, 49, 50, 51). Therefore, the mean of the sequence becomes 47.5, and the median is 49.5. Based on the formula of nonparametric skew, defined as $(\mu - \nu)/\sigma$, the skew is negative. Similarly, we can make the sequence positively skewed by adding a value far above the mean, which is probably a positive outlier, e.g. (49, 50, 51, 60), where the mean is 52.5, and the median is 50.5.

Relationship of Mean and Median

The skewness is not directly related to the relationship between the mean and median: a distribution with negative skew can have its mean greater than or less than the median, and likewise for positive skew.

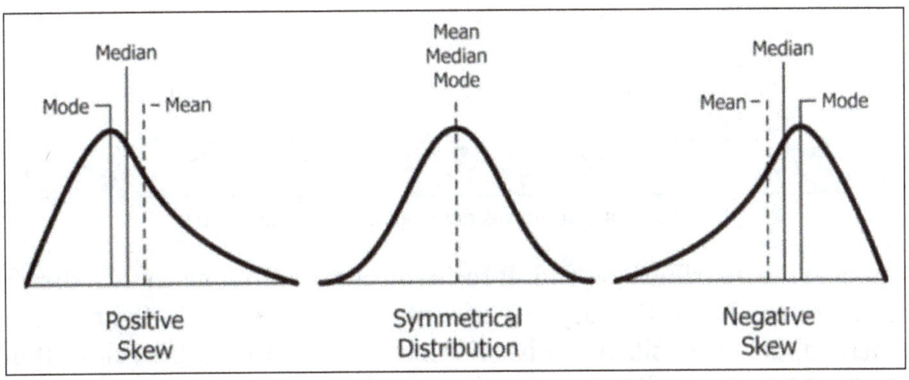

A general relationship of mean and median under differently skewed unimodal distribution.

In the older notion of nonparametric skew, defined as $(\mu - \nu)/\sigma$, where μ is the mean, ν is the median, and σ is the standard deviation, the skewness is defined in terms of this relationship: positive/right nonparametric skew means the mean is greater than (to the right of) the median, while negative/left nonparametric skew means the mean is less than (to the left of) the median. However, the modern definition of skewness and the traditional nonparametric definition do not always have the same sign: while they agree for some families of distributions, they differ in some of the cases, and conflating them is misleading.

If the distribution is symmetric, then the mean is equal to the median, and the distribution has zero skewness. If the distribution is both symmetric and unimodal, then the mean = median = mode. This is the case of a coin toss or the series 1,2,3,4,... Note, however, that the converse is not true in general, i.e. zero skewness does not imply that the mean is equal to the median.

A study points out:

> "Many textbooks teach a rule of thumb stating that the mean is right of the median under right skew, and left of the median under left skew. This rule fails with surprising frequency. It can fail in multimodal distributions, or in distributions where one tail is long but the other is heavy. Most commonly, though, the rule fails in discrete distributions where the areas to the left and right of the median are not equal. Such distributions not only contradict the textbook relationship between mean, median, and skew, they also contradict the textbook interpretation of the median".

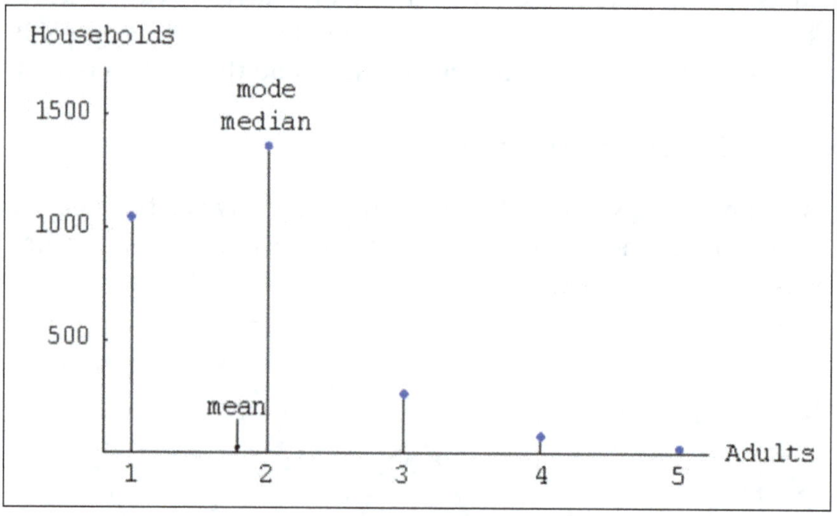

Distribution of adult residents across US households.

For example, in the distribution of adult residents across US households, the skew is to the right. However, due to the majority of cases is less or equal to the mode, which is also the median, the mean sits in the heavier left tail. As a result, the rule of thumb that the mean is right of the median under right skew failed.

Definition:

Pearson's Moment Coefficient of Skewness

The skewness of a random variable X is the third standardized moment $\tilde{\mu}_3$, defined as:

$$\tilde{\mu}_3 = \mathrm{E}\left[\left(\frac{X-\mu}{\sigma}\right)^3\right] = \frac{\mu_3}{\sigma^3} = \frac{\mathrm{E}\left[(X-\mu)^3\right]}{\left(\mathrm{E}\left[(X-\mu)^2\right]\right)^{3/2}} = \frac{\kappa_3}{\kappa_2^{3/2}}$$

where μ is the mean, σ is the standard deviation, E is the expectation operator, μ_3 is the third central moment, and κ_t are the t-th cumulants. It is sometimes referred to as Pearson's moment coefficient of skewness, or simply the moment coefficient of skewness, but should not be confused with Pearson's other skewness statistics. The last equality expresses skewness in terms of the ratio of the third cumulant κ_3 to the 1.5th power of the second cumulant κ_2. This is analogous to the definition of kurtosis as the fourth cumulant normalized by the square of the second cumulant. The skewness is also sometimes denoted Skew[X].

If σ is finite, μ is finite too and skewness can be expressed in terms of the non-central moment E[X^3] by expanding the previous formula,

$$\begin{aligned}\tilde{\mu}_3 &= \mathrm{E}\left[\left(\frac{X-\mu}{\sigma}\right)^3\right]\\ &= \frac{\mathrm{E}[X^3] - 3\mu\,\mathrm{E}[X^2] + 3\mu^2\,\mathrm{E}[X] - \mu^3}{\sigma^3}\\ &= \frac{\mathrm{E}[X^3] - 3\mu(\mathrm{E}[X^2] - \mu\,\mathrm{E}[X]) - \mu^3}{\sigma^3}\\ &= \frac{\mathrm{E}[X^3] - 3\mu\sigma^2 - \mu^3}{\sigma^3}.\end{aligned}$$

Examples:

Skewness can be infinite, as when,

$$\Pr[X > x] = x^{-2} \textit{ for } x > 1,\ \Pr[X < 1] = 0$$

where the third cumulants are infinite, or as when,

$$\Pr[X < x] = (1-x)^{-3}/2 \text{ for negative } x \text{ and}$$

$$\Pr[X > x] = (1+x)^{-3}/2 \text{ for positive } x.$$

where the third cumulant is undefined.

Examples of distributions with finite skewness include the following.

- A normal distribution and any other symmetric distribution with finite third moment has a skewness of 0.
- A half-normal distribution has a skewness just below 1.
- An exponential distribution has a skewness of 2.
- A lognormal distribution can have a skewness of any positive value, depending on its parameters.

Properties

Starting from a standard cumulant expansion around a normal distribution, one can show that:

$$\text{skewness} = 3\,(\text{mean} - \text{median})/\text{standard deviation} + O(\text{skewness}^2).$$

If Y is the sum of n independent and identically distributed random variables, all with the distribution of X, then the third cumulant of Y is n times that of X and the second cumulant of Y is n times that of X, so $\text{Skew}[Y] = \text{Skew}[X]/\sqrt{n}$. This shows that the skewness of the sum is smaller, as it approaches a Gaussian distribution in accordance with the central limit theorem. Note that the assumption that the variables be independent for the above formula is very important because it is possible even for the sum of two Gaussian variables to have a skewed distribution.

Sample Skewness

For a sample of n values, a natural method of moments estimator of the population skewness is:

$$b_1 = \frac{m_3}{s^3} = \frac{\frac{1}{n}\sum_{i=1}^{n}(x_i - \overline{x})^3}{\sqrt{\frac{1}{n-1}\sum_{i=1}^{n}(x_i - \overline{x})^2}^{\,3}} = \frac{\frac{1}{n}\sum_{i=1}^{n}(x_i - \overline{x})^3}{\left[\frac{1}{n-1}\sum_{i=1}^{n}(x_i - \overline{x})^2\right]^{3/2}},$$

where $\overline{x}$ is the sample mean, s is the sample standard deviation, and the numerator m_3 is the sample third central moment. This formula can be thought of as the average cubed deviation in the sample divided by the cubed sample standard deviation.

Another common definition of the *sample skewness* is:

$$G_1 = \frac{k_3}{k_2^{3/2}} = \frac{n^2}{(n-1)(n-2)}\,\frac{m_3}{s^3}$$

$$= \frac{\sqrt{n(n-1)}}{n-2} \frac{m_3}{m_2^{3/2}} = \frac{\sqrt{n(n-1)}}{n-2} \left[\frac{\frac{1}{n}\sum_{i=1}^{n}(x_i - \bar{x})^3}{\left(\frac{1}{n}\sum_{i=1}^{n}(x_i - \bar{x})^2\right)^{3/2}} \right],$$

where k_3 is the unique symmetric unbiased estimator of the third cumulant and $k_2 = s^2$ is the symmetric unbiased estimator of the second cumulant (i.e. the variance).

In general, the ratios b_1 and G_1 are both biased estimators of the population skewness γ_1; their expected values can even have the opposite sign from the true skewness. (For instance, a mixed distribution consisting of very thin Gaussians centred at −99, 0.5, and 2 with weights 0.01, 0.66, and 0.33 has a skewness of about −9.77, but in a sample of 3, G_1 has an expected value of about 0.32, since usually all three samples are in the positive-valued part of the distribution, which is skewed the other way.) Nevertheless, b_1 and G_1 each have obviously the correct expected value of zero for any symmetric distribution with a finite third moment, including a normal distribution.

Under the assumption that the underlying random variable X is normally distributed, it can be shown that $\sqrt{n}b_1 \xrightarrow{d} N(0,6)$, i.e., its distribution converges to a normal distribution with mean 0 and variance 6. The variance of the skewness of a random sample of size n from a normal distribution is:

$$\operatorname{var}(G_1) = \frac{6n(n-1)}{(n-2)(n+1)(n+3)}.$$

An approximate alternative is $6/n$, but this is inaccurate for small samples.

In normal samples, b_1 has the smaller variance of the two estimators, with:

$$\operatorname{var}(b_1) < \operatorname{var}\left(\frac{m_3}{m_2^{3/2}}\right) < \operatorname{var}(G_1),$$

where m_2 in the denominator is the (biased) sample second central moment.

The adjusted Fisher–Pearson standardized moment coefficient G_1 is the version found in Excel and several statistical packages including Minitab, SAS and SPSS.

Applications

Skewness is a descriptive statistic that can be used in conjunction with the histogram and the normal quantile plot to characterize the data or distribution.

Skewness indicates the direction and relative magnitude of a distribution's deviation from the normal distribution.

With pronounced skewness, standard statistical inference procedures such as a confidence interval for a mean will be not only incorrect, in the sense that the true coverage level will differ from the nominal (e.g., 95%) level, but they will also result in unequal error probabilities on each side.

Skewness can be used to obtain approximate probabilities and quantiles of distributions (such as value at risk in finance) via the Cornish-Fisher expansion.

Many models assume normal distribution; i.e., data are symmetric about the mean. The normal distribution has a skewness of zero. But in reality, data points may not be perfectly symmetric. So, an understanding of the skewness of the dataset indicates whether deviations from the mean are going to be positive or negative.

D'Agostino's K-squared test is a goodness-of-fit normality test based on sample skewness and sample kurtosis.

Other Measures of Skewness

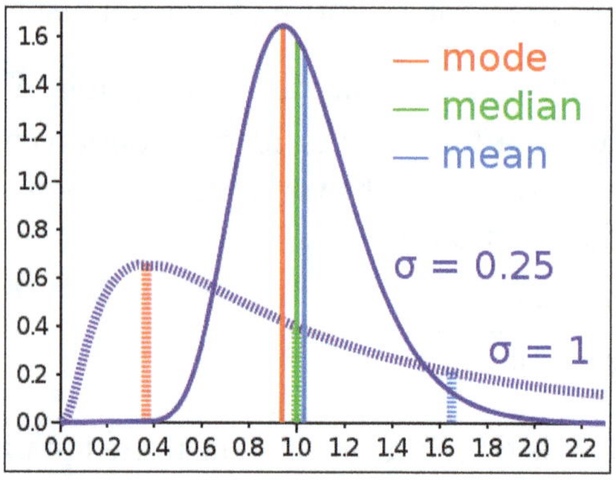

Comparison of mean, median and mode of two log-normal distributions with different skewnesses.

Other measures of skewness have been used, including simpler calculations suggested by Karl Pearson These other measures are:

Pearson's First Skewness Coefficient (Mode Skewness)

The Pearson mode skewness, or first skewness coefficient, is defined as:

(mean – mode)/standard deviation.

Pearson's Second Skewness Coefficient (Median Skewness)

The Pearson median skewness, or second skewness coefficient, is defined as:

3 (mean – median)/standard deviation.

Which is a simple multiple of the nonparametric skew.

Quartile-based Measures

Bowley's measure of skewness, also called Yule's coefficient is defined as:

$$B_1 = \frac{Q_3 + Q_1 - 2Q_2}{Q_3 - Q_1}$$

When writing it as $\dfrac{\dfrac{Q_3 + Q_1}{2} - Q_2}{\dfrac{Q_3 - Q_1}{2}}$, it is easier to see that the numerator is the average of the upper and lower quartiles (a measure of location) minus the median while the denominator is (Q3-Q1)/2 which (for symmetric distributions) is the MAD measure of dispersion.

Other names for this measure are Galton's measure of skewness, the Yule–Kendall index and the quartile skewness,

A more general formulation of a skewness function was described by Groeneveld, R. A. and Meeden, G:

$$\gamma(u) = \frac{F^{-1}(u) + F^{-1}(1-u) - 2F^{-1}(1/2)}{F^{-1}(u) - F^{-1}(1-u)}$$

where F is the cumulative distribution function. This leads to a corresponding overall measure of skewness defined as the supremum of this over the range $1/2 \leq u < 1$. Another measure can be obtained by integrating the numerator and denominator of this expression. The function $\gamma(u)$ satisfies $-1 \leq \gamma(u) \leq 1$ and is well defined without requiring the existence of any moments of the distribution. Quantile-based skewness measures are at first glance easy to interpret, but they often show significantly larger sample variations, than moment-based methods. This means that often samples from a symmetric distribution (like the uniform distribution) have a large quantile-based skewness, just by chance.

Bowley's measure of skewness is $\gamma(u)$ evaluated at $u = 3/4$. Kelley's measure of skewness uses $u = 0.1$.

Groeneveld and Meeden's Coefficient

Groeneveld and Meeden have suggested, as an alternative measure of skewness,

$$B_3 = \text{skew}(X) = \frac{(\mu - \nu)}{E(|X - \nu|)},$$

where μ is the mean, ν is the median, $|...|$ is the absolute value, and $E()$ is the expectation operator. This is closely related in form to Pearson's second skewness coefficient.

L-moments

Use of L-moments in place of moments provides a measure of skewness known as the L-skewness.

Distance Skewness

A value of skewness equal to zero does not imply that the probability distribution is symmetric. Thus there is a need for another measure of asymmetry that has this property: such a measure was introduced in 2000. It is called distance skewness and denoted by dSkew. If X is a random variable taking values in the d-dimensional Euclidean space, X has finite expectation, X' is an independent identically distributed copy of X, and $\|\cdot\|$ denotes the norm in the Euclidean space, then a simple *measure of asymmetry* with respect to location parameter θ is,

$$\mathrm{dSkew}(X) := 1 - \frac{\mathrm{E}\,\|X - X'\|}{\mathrm{E}\,\|X + X' - 2\theta\|} \quad \text{if}\Pr(X=\theta)\neq 1$$

and dSkew(X) := 0 for X = θ (with probability 1). Distance skewness is always between 0 and 1, equals 0 if and only if X is diagonally symmetric with respect to θ (X and 2θ−X have the same probability distribution) and equals 1 if and only if X is a constant c ($c \neq \theta$) with probability one. Thus there is a simple consistent statistical test of diagonal symmetry based on the sample distance skewness:

$$dSkew_n(X) := 1 - \frac{\sum_{i,j} \|x_i - x_j\|}{\sum_{i,j} \|x_i + x_j - 2\theta\|}.$$

Medcouple

The medcouple is a scale-invariant robust measure of skewness, with a breakdown point of 25%. It is the median of the values of the kernel function,

$$h(x_i, x_j) = \frac{(x_i - x_m) - (x_m - x_j)}{x_i - x_j}$$

taken over all couples (x_i, x_j) such that $x_i \geq x_m \geq x_j$, where x_m is the median of the sample $\{x_1, x_2, \ldots, x_n\}$.. It can be seen as the median of all possible quantile skewness measures.

STANDARD DEVIATION

In statistics, the standard deviation (SD, also represented by the lower case Greek letter sigma σ for the population standard deviation or the Latin letter s for the sample standard deviation) is a measure of the amount of variation or dispersion of a set of values. A low standard deviation indicates that the values tend to be close to the mean (also called the expected value) of the set, while a high standard deviation indicates that the values are spread out over a wider range.

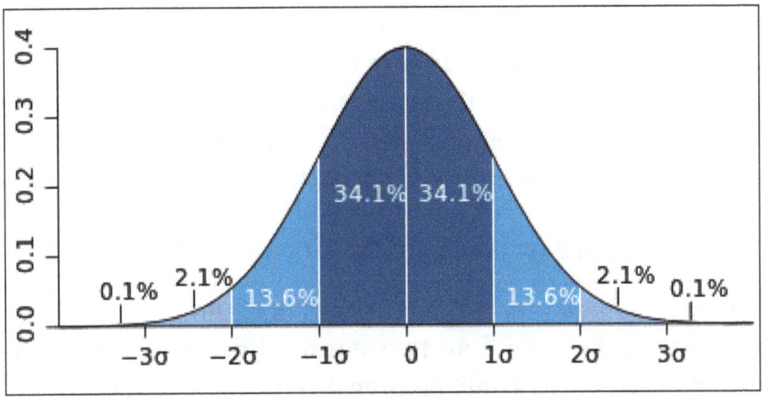

A plot of normal distribution (or bell-shaped curve) where each band has a width of 1 standard deviation.

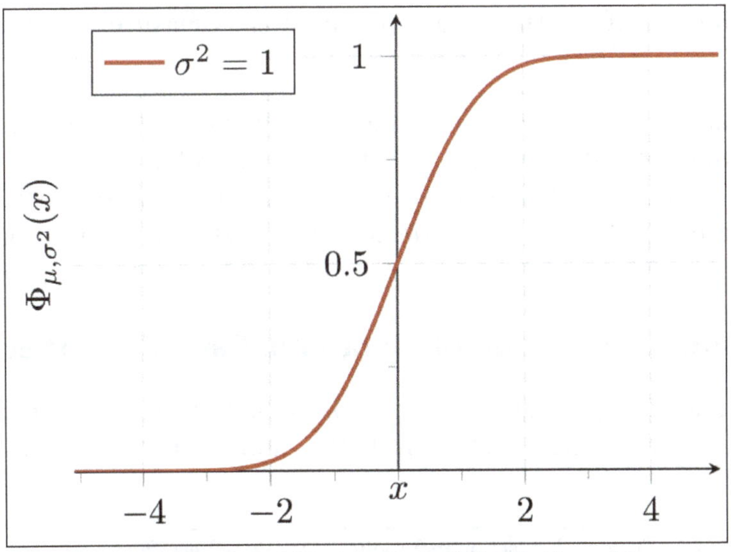

Cumulative probability of a normal distribution with expected value 0 and standard deviation 1.

The standard deviation of a random variable, statistical population, data set, or probability distribution is the square root of its variance. It is algebraically simpler, though in practice less robust, than the average absolute deviation. A useful property of the standard deviation is that, unlike the variance, it is expressed in the same units as the data.

In addition to expressing the variability of a population, the standard deviation is commonly used to measure confidence in statistical conclusions. For example, the margin of error in polling data is determined by calculating the expected standard deviation in the results if the same poll were to be conducted multiple times. This derivation of a standard deviation is often called the "standard error" of the estimate or "standard error of the mean" when referring to a mean. It is computed as the standard deviation of all the means that would be computed from that population if an infinite number of samples were drawn and a mean for each sample were computed.

The standard deviation of a population and the standard error of a statistic derived from that population (such as the mean) are quite different but related (related by the inverse of the square root of the number of observations). The reported margin of error of a poll is computed from the standard error of the mean (or alternatively from the product of the standard deviation of the population and the inverse of the square root of the sample size, which is the same thing) and is typically about twice the standard deviation—the half-width of a 95 percent confidence interval.

In science, many researchers report the standard deviation of experimental data, and by convention, only effects more than two standard deviations away from a null expectation are considered statistically significant—normal random error or variation in the measurements is in this way distinguished from likely genuine effects or associations. The standard deviation is also important in finance, where the standard deviation on the rate of return on an investment is a measure of the volatility of the investment.

When only a sample of data from a population is available, the term *standard deviation of the sample* or *sample standard deviation* can refer to either the above-mentioned quantity as applied to those data, or to a modified quantity that is an unbiased estimate of the *population standard deviation* (the standard deviation of the entire population).

Sample Standard Deviation of Metabolic Rate of Northern Fulmars

Logan gives the following example. Furness and Bryant measured the resting metabolic rate for 8 male and 6 female breeding northern fulmars. The table shows the Furness data set.

Furness data set on metabolic rates of northern fulmars			
Sex	Metabolic rate	Sex	Metabolic rate
Male	525.8	Female	727.7
Male	605.7	Female	1086.5
Male	843.3	Female	1091.0
Male	1195.5	Female	1361.3
Male	1945.6	Female	1490.5
Male	2135.6	Female	1956.1
Male	2308.7		
Male	2950.0		

The graph shows the metabolic rate for males and females. By visual inspection, it appears that the variability of the metabolic rate is greater for males than for females.

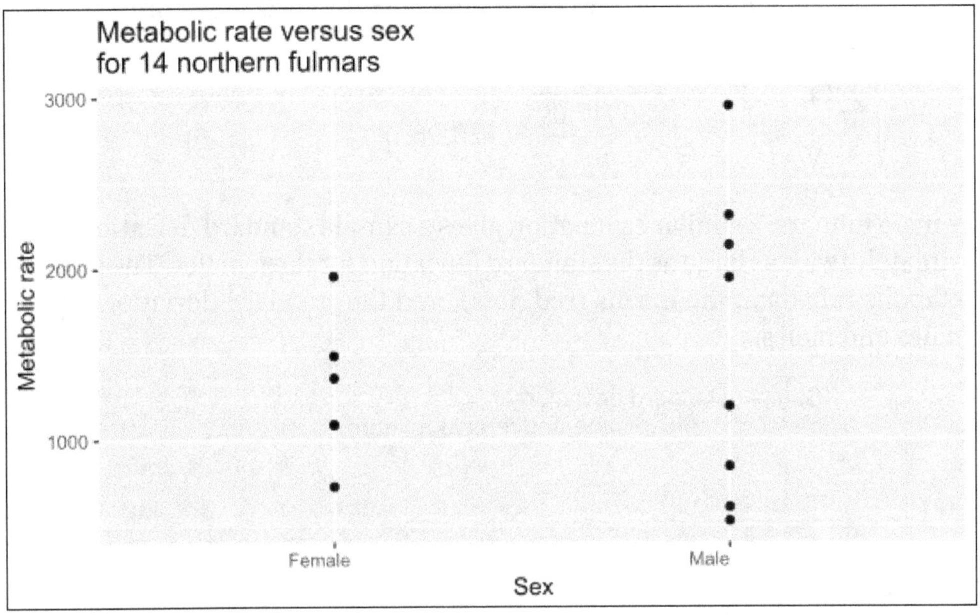

The sample standard deviation of the metabolic rate for the female fulmars is calculated as follows. The formula for the sample standard deviation is,

$$s = \sqrt{\frac{1}{N-1}\sum_{i=1}^{N}(x_i - \overline{x})^2},$$

where $\{x_1, x_2, \ldots, x_N\}$ are the observed values of the sample items, $\overline{x}$ is the mean value of these observations, and N is the number of observations in the sample.

In the sample standard deviation formula, for this example, the numerator is the sum of the squared deviation of each individual animal's metabolic rate from the mean metabolic rate. The table below shows the calculation of this sum of squared deviations for the female fulmars. For females, the sum of squared deviations is 886047.09, as shown in the table.

Sum of squares calculation for female fulmars					
Animal	Sex	Metabolic rate	Mean	Difference from mean	Squared difference from mean
1	Female	727.7	1285.5	-557.8	311140.84
2	Female	1086.5	1285.5	-199.0	39601.00
3	Female	1091.0	1285.5	-194.5	37830.25
4	Female	1361.3	1285.5	75.8	5745.64
5	Female	1490.5	1285.5	205.0	42025.00
6	Female	1956.1	1285.5	670.6	449704.36
Mean of metabolic rates			1285.5	Sum of squared differences	886047.09

The denominator in the sample standard deviation formula is $N - 1$, where N is the number of animals. In this example, there are $N = 6$ females, so the denominator is $6 - 1 = 5$. The sample standard deviation for the female fulmars is therefore,

$$s = \sqrt{\frac{\sum_{i=1}^{N}(x_i - \overline{x})^2}{N-1}} = \sqrt{\frac{886047.09}{5}} = 420.96.$$

For the male fulmars, a similar calculation gives a sample standard deviation of 894.37, approximately twice as large as the standard deviation for the females. The graph shows the metabolic rate data, the means (red dots), and the standard deviations (red lines) for females and males.

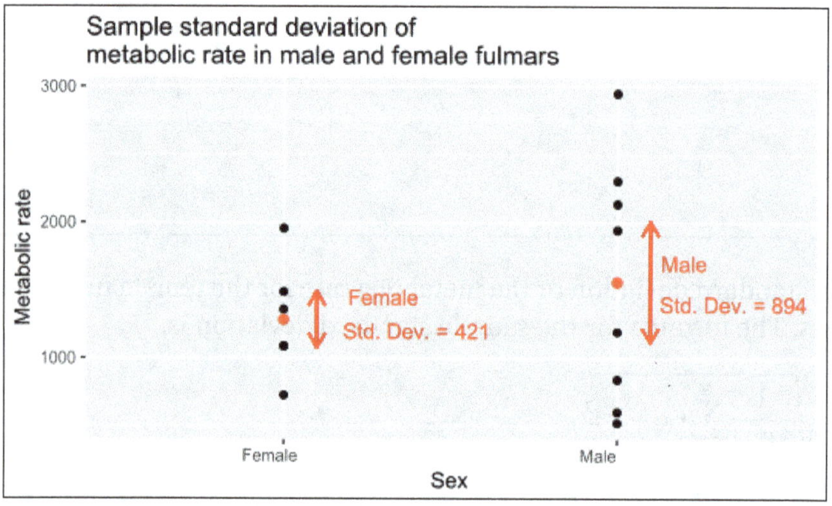

Use of the sample standard deviation implies that these 14 fulmars are a sample from a larger population of fulmars. If these 14 fulmars comprised the entire population (perhaps the last 14 surviving fulmars), then instead of the sample standard deviation, the calculation would use the population standard deviation. In the population standard deviation formula, the denominator is N instead of $N - 1$. It is rare that measurements can be taken for an entire population, so, by default, statistical computer programs calculate the sample standard deviation.

Population Standard Deviation of Grades of Eight Students

Suppose that the entire population of interest was eight students in a particular class. For a finite set of numbers, the population standard deviation is found by taking the square root of the average of the squared deviations of the values subtracted from their average value. The marks of a class of eight students (that is, a statistical population) are the following eight values:

$$2, 4, 4, 4, 5, 5, 7, 9.$$

These eight data points have the mean (average) of 5:

$$\mu = \frac{2+4+4+4+5+5+7+9}{8} = 5.$$

First, calculate the deviations of each data point from the mean, and square the result of each:

$$\begin{aligned} (2-5)^2 &= (-3)^2 = 9 & (5-5)^2 &= 0^2 = 0 \\ (4-5)^2 &= (-1)^2 = 1 & (5-5)^2 &= 0^2 = 0 \\ (4-5)^2 &= (-1)^2 = 1 & (7-5)^2 &= 2^2 = 4 \\ (4-5)^2 &= (-1)^2 = 1 & (9-5)^2 &= 4^2 = 16. \end{aligned}$$

The variance is the mean of these values:

$$\sigma^2 = \frac{9+1+1+1+0+0+4+16}{8} = 4.$$

and the *population* standard deviation is equal to the square root of the variance:

$$\sigma = \sqrt{4} = 2.$$

This formula is valid only if the eight values with which we began form the complete population. If the values instead were a random sample drawn from some large parent population (for example, they were 8 students randomly and independently chosen from a class of 2 million), then one often divides by 7 (which is $n - 1$) instead of 8 (which is n) in the denominator of the last formula. In that case the result of the original formula would be called the *sample* standard deviation. Dividing by $n - 1$ rather than by n gives an unbiased estimate of the variance of the larger parent population. This is known as *Bessel's correction*.

Standard Deviation of Average Height for Adult Men

If the population of interest is approximately normally distributed, the standard deviation provides information on the proportion of observations above or below certain values. For example, the average height for adult men in the United States is about 70 inches (177.8 cm), with a standard deviation of around 3 inches (7.62 cm). This means that most men (about 68%, assuming a normal distribution) have a height within 3 inches (7.62 cm) of the mean (67–73 inches (170.18–185.42 cm)) – one standard deviation – and almost all men (about 95%) have a height within 6 inches (15.24 cm) of the mean (64–76 inches (162.56–193.04 cm)) – two standard deviations. If the standard deviation were zero, then all men would be exactly 70 inches (177.8 cm) tall. If the standard deviation were 20 inches (50.8 cm), then men would have much more variable heights, with a typical range of about 50–90 inches (127–228.6 cm). Three standard deviations account for 99.7% of the sample population being studied, assuming the distribution is normal (bell-shaped).

Definition of Population Values

Let X be a random variable with mean value μ:

$$\mathrm{E}[X] = \mu.$$

Here the operator E denotes the average or expected value of X. Then the standard deviation of X is the quantity

$$\begin{aligned}\sigma &= \sqrt{\mathrm{E}[(X-\mu)^2]}\\ &= \sqrt{\mathrm{E}[X^2]+\mathrm{E}[-2\mu X]+\mathrm{E}[\mu^2]}\\ &= \sqrt{\mathrm{E}[X^2]-2\mu\mathrm{E}[X]+\mu^2}\\ &= \sqrt{\mathrm{E}[X^2]-2\mu^2+\mu^2}\\ &= \sqrt{\mathrm{E}[X^2]-\mu^2}\\ &= \sqrt{\mathrm{E}[X^2]-(\mathrm{E}[X])^2}\end{aligned}$$

(derived using the properties of expected value).

In other words, the standard deviation σ (sigma) is the square root of the variance of X; i.e., it is the square root of the average value of $(X - \mu)^2$.

The standard deviation of a (univariate) probability distribution is the same as that of a random variable having that distribution. Not all random variables have a standard deviation, since these expected values need not exist. For example, the standard deviation of a random variable that follows a Cauchy distribution is undefined because its expected value μ is undefined.

Discrete Random Variable

In the case where X takes random values from a finite data set $x_1, x_2, \ldots, x_N$, with each value having the same probability, the standard deviation is,

$$\sigma = \sqrt{\frac{1}{N}\left[(x_1-\mu)^2+(x_2-\mu)^2+\cdots+(x_N-\mu)^2\right]}, \text{ where } \mu = \frac{1}{N}(x_1+\cdots+x_N),$$

or, using summation notation,

$$\sigma = \sqrt{\frac{1}{N}\sum_{i=1}^{N}(x_i-\mu)^2}, \text{ where } \mu = \frac{1}{N}\sum_{i=1}^{N}x_i.$$

If, instead of having equal probabilities, the values have different probabilities, let x_1 have probability p_1, x_2 have probability p_2, ..., x_N have probability p_N. In this case, the standard deviation will be,

$$\sigma = \sqrt{\sum_{i=1}^{N} p_i (x_i - \mu)^2}, \text{ where } \mu = \sum_{i=1}^{N} p_i x_i.$$

Continuous Random Variable

The standard deviation of a continuous real-valued random variable X with probability density function $p(x)$ is,

$$\sigma = \sqrt{\int_{\mathbf{X}} (x - \mu)^2 \, p(x) \mathrm{d}x}, \text{ where } \mu = \int_{\mathbf{X}} x \, p(x) \mathrm{d}x,$$

and where the integrals are definite integrals taken for x ranging over the set of possible values of the random variable X.

In the case of a parametric family of distributions, the standard deviation can be expressed in terms of the parameters. For example, in the case of the log-normal distribution with parameters μ and σ^2, the standard deviation is,

$$\sqrt{(e^{\sigma^2} - 1) e^{2\mu + \sigma^2}}.$$

Estimation

One can find the standard deviation of an entire population in cases (such as standardized testing) where every member of a population is sampled. In cases where that cannot be done, the standard deviation σ is estimated by examining a random sample taken from the population and computing a statistic of the sample, which is used as an estimate of the population standard deviation. Such a statistic is called an estimator, and the estimator (or the value of the estimator, namely the estimate) is called a sample standard deviation, and is denoted by s (possibly with modifiers).

Unlike in the case of estimating the population mean, for which the sample mean is a simple estimator with many desirable properties (unbiased, efficient, maximum likelihood), there is no single estimator for the standard deviation with all these properties, and unbiased estimation of standard deviation is a very technically involved problem. Most often, the standard deviation is estimated using the *corrected sample standard deviation* (using $N - 1$), defined below, and this is often referred to as the "sample standard deviation", without qualifiers. However, other estimators are better in other respects: the uncorrected estimator (using N) yields lower mean squared error, while using $N - 1.5$ (for the normal distribution) almost completely eliminates bias.

Uncorrected Sample Standard Deviation

The formula for the population standard deviation (of a finite population) can be applied to the sample, using the size of the sample as the size of the population (though the actual population size from which the sample is drawn may be much larger). This estimator, denoted by s_N, is known as the uncorrected sample standard deviation, or sometimes the standard deviation of the sample (considered as the entire population), and is defined as follows:

$$s_N = \sqrt{\frac{1}{N}\sum_{i=1}^{N}(x_i - \overline{x})^2},$$

where $\{x_1, x_2, \ldots, x_N\}$ are the observed values of the sample items and $\overline{x}$ is the mean value of these observations, while the denominator N stands for the size of the sample: this is the square root of the sample variance, which is the average of the squared deviations about the sample mean.

This is a consistent estimator (it converges in probability to the population value as the number of samples goes to infinity), and is the maximum-likelihood estimate when the population is normally distributed. However, this is a biased estimator, as the estimates are generally too low. The bias decreases as sample size grows, dropping off as $1/N$, and thus is most significant for small or moderate sample sizes; for $N > 75$ the bias is below 1%. Thus for very large sample sizes, the uncorrected sample standard deviation is generally acceptable. This estimator also has a uniformly smaller mean squared error than the corrected sample standard deviation.

Corrected Sample Standard Deviation

If the *biased sample variance* (the second central moment of the sample, which is a downward-biased estimate of the population variance) is used to compute an estimate of the population's standard deviation, the result is,

$$s_N = \sqrt{\frac{1}{N}\sum_{i=1}^{N}(x_i - \overline{x})^2}.$$

Here taking the square root introduces further downward bias, by Jensen's inequality, due to the square root's being a concave function. The bias in the variance is easily corrected, but the bias from the square root is more difficult to correct, and depends on the distribution in question.

An unbiased estimator for the *variance* is given by applying Bessel's correction, using $N - 1$ instead of N to yield the *unbiased sample variance,* denoted s^2:

$$s^2 = \frac{1}{N-1}\sum_{i=1}^{N}(x_i - \overline{x})^2.$$

This estimator is unbiased if the variance exists and the sample values are drawn independently with replacement. $N - 1$ corresponds to the number of degrees of freedom in the vector of deviations from the mean,

$$(x_1 - \bar{x}, \ldots, x_n - \bar{x}).$$

Taking square roots reintroduces bias (because the square root is a nonlinear function, which does not commute with the expectation), yielding the *corrected sample standard deviation,* denoted by *s*:

$$s = \sqrt{\frac{1}{N-1}\sum_{i=1}^{N}(x_i - \bar{x})^2}.$$

As explained above, while s^2 is an unbiased estimator for the population variance, *s* is still a biased estimator for the population standard deviation, though markedly less biased than the uncorrected sample standard deviation. This estimator is commonly used and generally known simply as the "sample standard deviation". The bias may still be large for small samples (*N* less than 10). As sample size increases, the amount of bias decreases. We obtain more information and the difference between $\frac{1}{N}$ and $\frac{1}{N-1}$ becomes smaller.

Unbiased Sample Standard Deviation

For unbiased estimation of standard deviation, there is no formula that works across all distributions, unlike for mean and variance. Instead, *s* is used as a basis, and is scaled by a correction factor to produce an unbiased estimate. For the normal distribution, an unbiased estimator is given by s/c_4, where the correction factor (which depends on *N*) is given in terms of the Gamma function, and equals:

$$c_4(N)=\sqrt{\frac{2}{N-1}}\frac{\Gamma\left(\frac{N}{2}\right)}{\Gamma\left(\frac{N-1}{2}\right)}.$$

This arises because the sampling distribution of the sample standard deviation follows a (scaled) chi distribution, and the correction factor is the mean of the chi distribution.

An approximation can be given by replacing $N - 1$ with $N - 1.5$, yielding:

$$\hat{\sigma} = \sqrt{\frac{1}{N-1.5}\sum_{i=1}^{N}(x_i - \bar{x})^2},$$

The error in this approximation decays quadratically (as $1/N^2$), and it is suited for all but the smallest samples or highest precision: for N = 3 the bias is equal to 1.3%, and for N = 9 the bias is already less than 0.1%. A more accurate approximation is to replace $N-1.5$ above with $N-1.5+1/(8(N-1))$.

For other distributions, the correct formula depends on the distribution, but a rule of thumb is to use the further refinement of the approximation:

$$\hat{\sigma}=\sqrt{\frac{1}{N-1.5-\frac{1}{4}\gamma_2}\sum_{i=1}^{N}(x_i-\overline{x})^2},$$

where γ_2 denotes the population excess kurtosis. The excess kurtosis may be either known beforehand for certain distributions, or estimated from the data.

Confidence Interval of a Sampled Standard Deviation

The standard deviation we obtain by sampling a distribution is itself not absolutely accurate, both for mathematical reasons (explained here by the confidence interval) and for practical reasons of measurement (measurement error). The mathematical effect can be described by the confidence interval or CI. To show how a larger sample will make the confidence interval narrower, consider the following examples: A small population of N = 2 has only 1 degree of freedom for estimating the standard deviation. The result is that a 95% CI of the SD runs from 0.45 × SD to 31.9 × SD; the factors here are as follows:

$$\Pr\left(q_{\frac{\alpha}{2}}<k\frac{s^2}{\sigma^2}<q_{1-\frac{\alpha}{2}}\right)=1-\alpha,$$

where q_p is the p-th quantile of the chi-square distribution with k degrees of freedom, and $1-\alpha$ is the confidence level. This is equivalent to the following:

$$\Pr\left(k\frac{s^2}{q_{1-\frac{\alpha}{2}}}<\sigma^2<k\frac{s^2}{q_{\frac{\alpha}{2}}}\right)=1-\alpha.$$

With k = 1, $q_{0.025}=0.000982$ and $q_{0.975}=5.024$. The reciprocals of the square roots of these two numbers give us the factors 0.45 and 31.9 given above.

A larger population of N = 10 has 9 degrees of freedom for estimating the standard deviation. The same computations as above give us in this case a 95% CI running from 0.69 × SD to 1.83 × SD. So even with a sample population of 10, the actual SD can still be almost a factor 2 higher than the sampled SD. For a sample population N=100, this is down to 0.88 × SD to 1.16 × SD. To be more certain that the sampled SD is close to the actual SD we need to sample a large number of points.

These same formulae can be used to obtain confidence intervals on the variance of residuals from a least squares fit under standard normal theory, where k is now the number of degrees of freedom for error.

Bounds on Standard Deviation

For a set of N > 4 data spanning a range of values R, an upper bound on the standard deviation s is given by s = 0.6R. An estimate of the standard deviation for N > 100 data taken to be approximately normal follows from the heuristic that 95% of the area under the normal curve lies roughly two standard deviations to either side of the mean, so that, with 95% probability the total range of values R represents four standard deviations so that s ≈ R/4. This so-called range rule is useful in sample size estimation, as the range of possible values is easier to estimate than the standard deviation. Other divisors K(N) of the range such that s ≈ R/K(N) are available for other values of N and for non-normal distributions.

Identities and Mathematical Properties

The standard deviation is invariant under changes in location, and scales directly with the scale of the random variable. Thus, for a constant c and random variables X and Y:

$$\sigma(c) = 0$$

$$\sigma(X + c) = \sigma(X),$$

$$\sigma(cX) = |c|\,\sigma(X).$$

The standard deviation of the sum of two random variables can be related to their individual standard deviations and the covariance between them:

$$\sigma(X + Y) = \sqrt{\mathrm{var}(X) + \mathrm{var}(Y) + 2\mathrm{cov}(X, Y)}.$$

where $\mathrm{var}=\sigma^2$ and cov stand for variance and covariance, respectively.

The calculation of the sum of squared deviations can be related to moments calculated directly from the data. In the following formula, the letter E is interpreted to mean expected value, i.e., mean.

$$\sigma(X) = \sqrt{\mathrm{E}[(X - \mathrm{E}[X])^2]} = \sqrt{\mathrm{E}[X^2] - (\mathrm{E}[X])^2}.$$

The sample standard deviation can be computed as:

$$\mathrm{s}(X) = \sqrt{\frac{N}{N-1}}\sqrt{\mathrm{E}[(X - \mathrm{E}[X])^2]}.$$

For a finite population with equal probabilities at all points, we have:

$$\sqrt{\frac{1}{N}\sum_{i=1}^{N}(x_i-\overline{x})^2}=\sqrt{\frac{1}{N}\left(\sum_{i=1}^{N}x_i^2\right)-(\overline{x})^2}=\sqrt{\left(\frac{1}{N}\sum_{i=1}^{N}x_i^2\right)-\left(\frac{1}{N}\sum_{i=1}^{N}x_i\right)^2}.$$

This means that the standard deviation is equal to the square root of the difference between the average of the squares of the values and the square of the average value.

Interpretation and Application

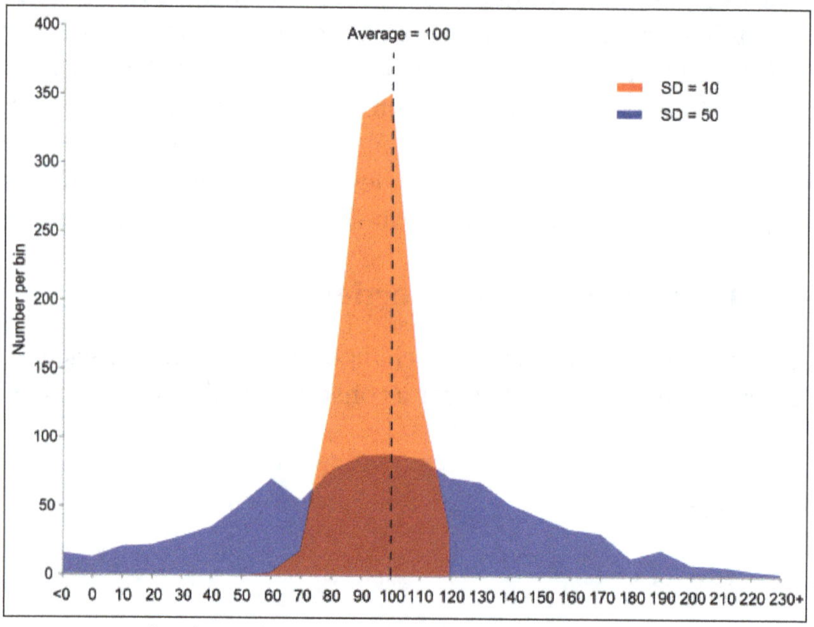

Example of samples from two populations with the same mean but different standard deviations. Red population has mean 100 and SD 10; blue population has mean 100 and SD 50.

A large standard deviation indicates that the data points can spread far from the mean and a small standard deviation indicates that they are clustered closely around the mean.

For example, each of the three populations {0, 0, 14, 14}, {0, 6, 8, 14} and {6, 6, 8, 8} has a mean of 7. Their standard deviations are 7, 5, and 1, respectively. The third population has a much smaller standard deviation than the other two because its values are all close to 7. It will have the same units as the data points themselves. If, for instance, the data set {0, 6, 8, 14} represents the ages of a population of four siblings in years, the standard deviation is 5 years. As another example, the population {1000, 1006, 1008, 1014} may represent the distances traveled by four athletes, measured in meters. It has a mean of 1007 meters, and a standard deviation of 5 meters.

Standard deviation may serve as a measure of uncertainty. In physical science, for example, the reported standard deviation of a group of repeated measurements gives the precision of those measurements. When deciding whether measurements agree with a theoretical prediction, the standard deviation of those measurements is of crucial importance: If the mean of the measurements is too far away from the prediction (with the distance measured in standard deviations), then the theory being tested probably needs to be revised. This makes sense since they fall outside the range of values that could reasonably be expected to occur, if the prediction were correct and the standard deviation appropriately quantified.

While the standard deviation does measure how far typical values tend to be from the mean, other measures are available. An example is the mean absolute deviation, which might be considered a more direct measure of average distance, compared to the root mean square distance inherent in the standard deviation.

Application Examples

The practical value of understanding the standard deviation of a set of values is in appreciating how much variation there is from the average (mean).

Experiment, Industrial and Hypothesis Testing

Standard deviation is often used to compare real-world data against a model to test the model. For example, in industrial applications the weight of products coming off a production line may need to comply with a legally required value. By weighing some fraction of the products an average weight can be found, which will always be slightly different from the long-term average. By using standard deviations, a minimum and maximum value can be calculated that the averaged weight will be within some very high percentage of the time (99.9% or more). If it falls outside the range then the production process may need to be corrected. Statistical tests such as these are particularly important when the testing is relatively expensive. For example, if the product needs to be opened and drained and weighed, or if the product was otherwise used up by the test.

In experimental science, a theoretical model of reality is used. Particle physics conventionally uses a standard of "5 sigma" for the declaration of a discovery. A five-sigma level translates to one chance in 3.5 million that a random fluctuation would yield the result. This level of certainty was required in order to assert that a particle consistent with the Higgs boson had been discovered in two independent experiments at CERN, and this was also the significance level leading to the declaration of the first detection of gravitational waves.

Weather

As a simple example, consider the average daily maximum temperatures for two cities, one inland and one on the coast. It is helpful to understand that the range of daily maximum temperatures for cities near the coast is smaller than for cities inland. Thus, while these two cities may each have the same average maximum temperature, the standard deviation of the daily maximum temperature for the coastal city will be less than that of the inland city as, on any particular day, the actual maximum temperature is more likely to be farther from the average maximum temperature for the inland city than for the coastal one.

Finance

In finance, standard deviation is often used as a measure of the risk associated with price-fluctuations of a given asset (stocks, bonds, property, etc.), or the risk of a portfolio of assets (actively managed mutual funds, index mutual funds, or ETFs). Risk is an important factor in determining how to efficiently manage a portfolio of investments because it determines the variation in returns on the asset and/or portfolio and gives investors a mathematical basis for investment decisions (known as mean-variance optimization). The fundamental concept of risk is that as it increases, the expected return on an investment should increase as well, an increase known as the risk premium. In other words, investors should expect a higher return on an investment when that investment carries a higher level of risk or uncertainty. When evaluating investments, investors should estimate both the expected return and the uncertainty of future returns. Standard deviation provides a quantified estimate of the uncertainty of future returns.

For example, assume an investor had to choose between two stocks. Stock A over the past 20 years had an average return of 10 percent, with a standard deviation of 20 percentage points (pp) and Stock B, over the same period, had average returns of 12 percent but a higher standard deviation of 30 pp. On the basis of risk and return, an investor may decide that Stock A is the safer choice, because Stock B's additional two percentage points of return is not worth the additional 10 pp standard deviation (greater risk or uncertainty of the expected return). Stock B is likely to fall short of the initial investment (but also to exceed the initial investment) more often than Stock A under the same circumstances, and is estimated to return only two percent more on average. In this example, Stock A is expected to earn about 10 percent, plus or minus 20 pp (a range of 30 percent to −10 percent), about two-thirds of the future year returns. When considering more extreme possible returns or outcomes in future, an investor should expect results of as much as 10 percent plus or minus 60 pp, or a range from 70 percent to −50 percent, which includes outcomes for three standard deviations from the average return (about 99.7 percent of probable returns).

Calculating the average (or arithmetic mean) of the return of a security over a given period will generate the expected return of the asset. For each period, subtracting the expected return from the actual return results in the difference from the mean. Squaring the difference in each period and taking the average gives the overall variance of the return of the asset. The larger the variance, the greater risk the security carries. Finding the square root of this variance will give the standard deviation of the investment tool in question.

Population standard deviation is used to set the width of Bollinger Bands, a widely adopted technical analysis tool. For example, the upper Bollinger Band is given as $\bar{x} + n\sigma_x$. The most commonly used value for n is 2; there is about a five percent chance of going outside, assuming a normal distribution of returns.

Financial time series are known to be non-stationary series, whereas the statistical calculations above, such as standard deviation, apply only to stationary series. To apply the above statistical tools to non-stationary series, the series first must be transformed to a stationary series, enabling use of statistical tools that now have a valid basis from which to work.

Geometric Interpretation

To gain some geometric insights and clarification, we will start with a population of three values, x_1, x_2, x_3. This defines a point $P = (x_1, x_2, x_3)$ in R^3. Consider the line $L = \{(r, r, r) : r \in R\}$. This is the "main diagonal" going through the origin. If our three given values were all equal, then the standard deviation would be zero and P would lie on L. So it is not unreasonable to assume that the standard deviation is related to the *distance* of P to L. That is indeed the case. To move orthogonally from L to the point P, one begins at the point:

$$M = (\bar{x}, \bar{x}, \bar{x})$$

whose coordinates are the mean of the values we started out with.

M is on L therefore $M = (\ell, \ell, \ell)$ for some $\ell \in \mathbb{R}$.

The line L is to be orthogonal to the vector from M to P. Therefore:

$$\begin{aligned}
L \cdot (P - M) &= 0 \\
(r, r, r) \cdot (x_1 - \ell, x_2 - \ell, x_3 - \ell) &= 0 \\
r(x_1 - \ell + x_2 - \ell + x_3 - \ell) &= 0 \\
r\left(\sum_i x_i - 3\ell\right) &= 0 \\
\sum_i x_i - 3\ell &= 0 \\
\frac{1}{3}\sum_i x_i &= \ell \\
\bar{x} &= \ell
\end{aligned}$$

A little algebra shows that the distance between P and M (which is the same as the orthogonal distance between P and the line L) $\sqrt{\sum_i (x_i - \bar{x})^2}$ is equal to the standard deviation of the vector (x_1, x_2, x_3), multiplied by the square root of the number of dimensions of the vector (3 in this case).

Chebyshev's Inequality

An observation is rarely more than a few standard deviations away from the mean. Chebyshev's inequality ensures that, for all distributions for which the standard deviation is defined, the amount of data within a number of standard deviations of the mean is at least as much as given in the following table.

Distance from mean	Minimum population
$\sqrt{2}\sigma$	50%
2σ	75%
3σ	89%
4σ	94%
5σ	96%
6σ	97%
$k\sigma$	$1-\frac{1}{k^2}$
$\frac{1}{\sqrt{1-\ell}}\sigma$	ℓ

Rules for Normally Distributed Data

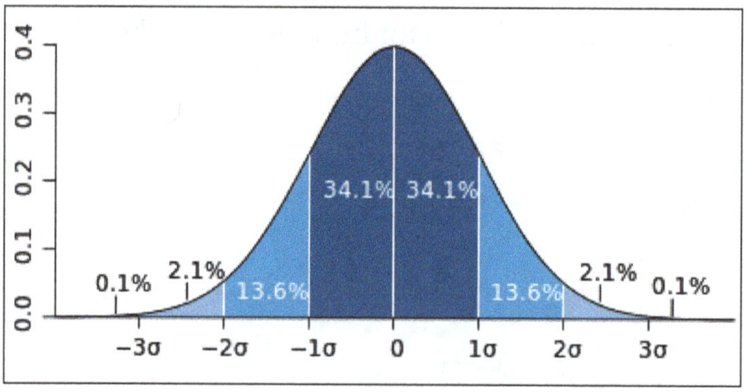

Dark blue is one standard deviation on either side of the mean. For the normal distribution, this accounts for 68.27 percent of the set; while two standard deviations from the mean (medium and dark blue) account for 95.45 percent; three standard deviations (light, medium, and dark blue) account for 99.73 percent; and four standard deviations account for 99.994 percent. The two points of the curve that are one standard deviation from the mean are also the inflection points.

The central limit theorem states that the distribution of an average of many independent, identically distributed random variables tends toward the famous bell-shaped normal distribution with a probability density function of:

$$f(x;\mu,\sigma^2)=\frac{1}{\sigma\sqrt{2\pi}}e^{-\frac{1}{2}\left(\frac{x-\mu}{\sigma}\right)^2}$$

where μ is the expected value of the random variables, σ equals their distribution's standard deviation divided by $n^{1/2}$, and n is the number of random variables. The standard deviation therefore is simply a scaling variable that adjusts how broad the curve will be, though it also appears in the normalizing constant.

If a data distribution is approximately normal, then the proportion of data values within z standard deviations of the mean is defined by:

$$\text{Proportion} = erf\left(\frac{z}{\sqrt{2}}\right)$$

where erf is the error function. The proportion that is less than or equal to a number, x, is given by the cumulative distribution function:

$$\text{Proportion} \leq x = \frac{1}{2}\left[1+\text{erf}\left(\frac{x-\mu}{\sigma\sqrt{2}}\right)\right]=\frac{1}{2}\left[1+\text{erf}\left(\frac{z}{\sqrt{2}}\right)\right].$$

If a data distribution is approximately normal then about 68 percent of the data values are within one standard deviation of the mean (mathematically, $\mu \pm \sigma$, where μ is the arithmetic mean), about 95 percent are within two standard deviations ($\mu \pm 2\sigma$), and about 99.7 percent lie within three standard deviations ($\mu \pm 3\sigma$). This is known as the *68-95-99.7 rule*, or *the empirical rule.*

For various values of z, the percentage of values expected to lie in and outside the symmetric interval, CI = ($-z\sigma$, $z\sigma$), are as follows:

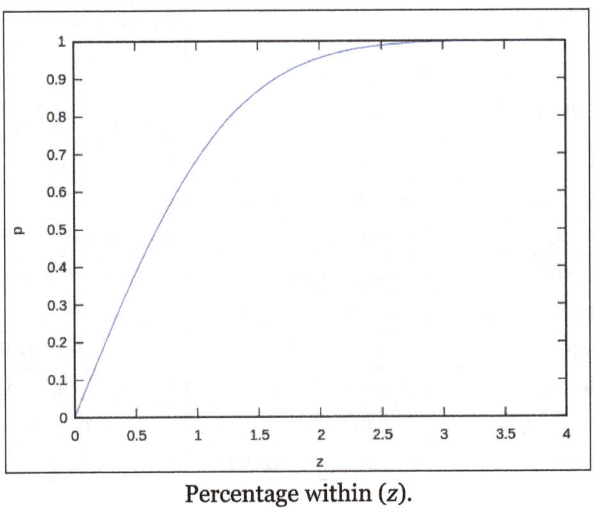

Percentage within (z).

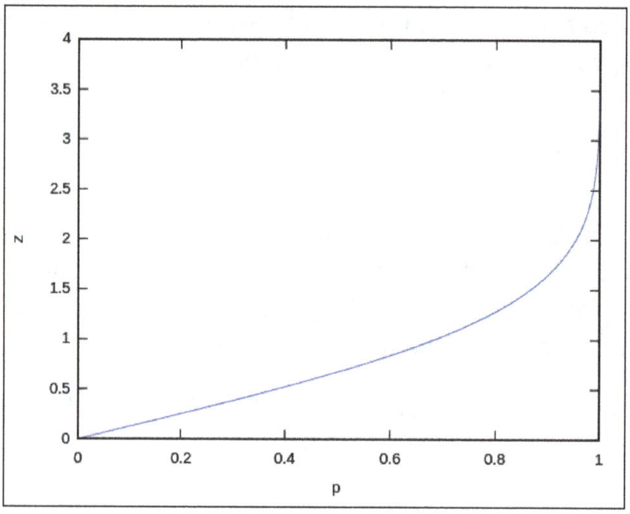

z (Percentage within).

Confidence interval		Proportion without	
	Percentage	Percentage	Fraction
0.318 639σ	25%	75%	3 / 4
0.674490σ	50%	50%	1 / 2
0.994458σ	68%	32%	1 / 3.125
1σ	68.2689492%	31.7310508%	1 / 3.1514872
1.281552σ	80%	20%	1 / 5
1.644854σ	90%	10%	1 / 10
1.959964σ	95%	5%	1 / 20
2σ	95.4499736%	4.5500264%	1 / 21.977895
2.575829σ	99%	1%	1 / 100
3σ	99.7300204%	0.2699796%	1 / 370.398
3.290527σ	99.9%	0.1%	1 / 1000
3.890592σ	99.99%	0.01%	1 / 10000
4σ	99.993666%	0.006334%	1 / 15787
4.417173σ	99.999%	0.001%	1 / 100000
4.5σ	99.9993204653751%	0.0006795346249%	1 / 147159.5358 3.4 / 1000000 (on each side of mean)
4.891638σ	99.9999%	0.0001%	1 / 1000000
5σ	99.9999426697%	0.0000573303%	1 / 1744278
5.326724σ	99.99999%	0.00001%	1 / 10000000
5.730729σ	99.999999%	0.000001%	1 / 100000000
6σ	99.9999998027%	0.0000001973%	1 / 506797346
6.109410σ	99.9999999%	0.0000001%	1 / 1000000000
6.466951σ	99.99999999%	0.00000001%	1 / 10000000000
6.806502σ	99.999999999%	0.000000001%	1 / 100000000000
7σ	99.9999999997440%	0.000000000256%	1 / 390682215445

Relationship between Standard Deviation and Mean

The mean and the standard deviation of a set of data are descriptive statistics usually reported together. In a certain sense, the standard deviation is a "natural" measure of statistical dispersion if the center of the data is measured about the mean. This is because the standard deviation from the mean is smaller than from any other point. The precise statement is the following: suppose $x_1, ..., x_n$ are real numbers and define the function:

$$\sigma(r) = \sqrt{\frac{1}{N-1}\sum_{i=1}^{N}(x_i - r)^2}.$$

Using calculus or by completing the square, it is possible to show that $\sigma(r)$ has a unique minimum at the mean:

$$r = \bar{x}.$$

Variability can also be measured by the coefficient of variation, which is the ratio of the standard deviation to the mean. It is a dimensionless number.

Standard Deviation of the Mean

Often, we want some information about the precision of the mean we obtained. We can obtain this by determining the standard deviation of the sampled mean. Assuming statistical independence of the values in the sample, the standard deviation of the mean is related to the standard deviation of the distribution by:

$$\sigma_{\text{mean}} = \frac{1}{\sqrt{N}}\sigma$$

where N is the number of observations in the sample used to estimate the mean. This can easily be proven with:

$$\text{var}(X) \equiv \sigma_X^2$$
$$\text{var}(X_1 + X_2) \equiv \text{var}(X_1) + \text{var}(X_2)$$

(Statistical Independence is assumed).

$$\text{var}(cX_1) \equiv c^2 \text{var}(X_1)$$

hence,

$$\text{var(mean)} = \text{var}\left(\frac{1}{N}\sum_{i=1}^{N} X_i\right) = \frac{1}{N^2}\text{var}\left(\sum_{i=1}^{N} X_i\right)$$
$$= \frac{1}{N^2}\sum_{i=1}^{N}\text{var}(X_i) = \frac{N}{N^2}\text{var}(X) = \frac{1}{N}\text{var}(X).$$

Resulting in:

$$\sigma_{\text{mean}} = \frac{\sigma}{\sqrt{N}}.$$

It should be emphasized that in order to estimate the standard deviation of the mean σ_{mean} it is necessary to know the standard deviation of the entire population σ beforehand. However, in most applications this parameter is unknown. For example, if a series of 10 measurements of a previously unknown quantity is performed in a laboratory, it is possible to calculate the resulting sample mean and sample standard deviation, but it is impossible to calculate the standard deviation of the mean.

Rapid Calculation Methods

The following two formulas can represent a running (repeatedly updated) standard deviation. A set of two power sums s_1 and s_2 are computed over a set of N values of x, denoted as $x_1, ..., x_N$:

$$s_j = \sum_{k=1}^{N} x_k^j.$$

Given the results of these running summations, the values N, s_1, s_2 can be used at any time to compute the *current* value of the running standard deviation:

$$\sigma = \frac{\sqrt{Ns_2 - s_1^2}}{N}$$

Where N, as mentioned above, is the size of the set of values (or can also be regarded as s_0).

Similarly for sample standard deviation,

$$= \sqrt{\frac{Ns_2 - s_1^2}{N(N-1)}}.$$

In a computer implementation, as the three s_j sums become large, we need to consider round-off error, arithmetic overflow, and arithmetic underflow. The method below calculates the running sums method with reduced rounding errors. This is a "one pass" algorithm for calculating variance of n samples without the need to store prior data during the calculation. Applying this method to a time series will result in successive values of standard deviation corresponding to n data points as n grows larger with each new sample, rather than a constant-width sliding window calculation.

For $k = 1, \ldots, n$:

$$A_0 = 0$$

$$A_k = A_{k-1} + \frac{x_k - A_{k-1}}{k}$$

where A is the mean value,

$$Q_0 = 0$$

$$Q_k = Q_{k-1} + \frac{k-1}{k}(x_k - A_{k-1})^2 = Q_{k-1} + (x_k - A_{k-1})(x_k - A_k)$$

Note: $Q_1 = 0$ since $k-1=0$ or $x_1 = A_1$

Sample variance,

$$s_n^2 = \frac{Q_n}{n-1}$$

Population variance,

$$\sigma_n^2 = \frac{Q_n}{n}$$

Weighted Calculation

When the values x_i are weighted with unequal weights w_i, the power sums s_0, s_1, s_2 are each computed as:

$$s_j = \sum_{k=1}^{N} w_k x_k^j.$$

And the standard deviation equations remain unchanged. s_0 is now the sum of the weights and not the number of samples N.

The incremental method with reduced rounding errors can also be applied, with some additional complexity.

A running sum of weights must be computed for each k from 1 to n:

$$W_0 = 0$$
$$W_k = W_{k-1} + w_k$$

and places where $1/n$ is used above must be replaced by w_i/W_n:

$$A_0 = 0$$

$$A_k = A_{k-1} + \frac{w_k}{W_k}(x_k - A_{k-1})$$

$$Q_0 = 0$$

$$Q_k = Q_{k-1} + \frac{w_k W_{k-1}}{W_k}(x_k - A_{k-1})^2 = Q_{k-1} + w_k(x_k - A_{k-1})(x_k - A_k)$$

In the final division,

$$\sigma_n^2 = \frac{Q_n}{W_n}$$

and

$$s_n^2 = \frac{Q_n}{W_n - 1},$$

or

$$s_n^2 = \frac{n'}{n'-1}\sigma_n^2,$$

where n is the total number of elements, and n' is the number of elements with non-zero weights. The above formulas become equal to the simpler formulas given above if weights are taken as equal to one.

QUANTILE

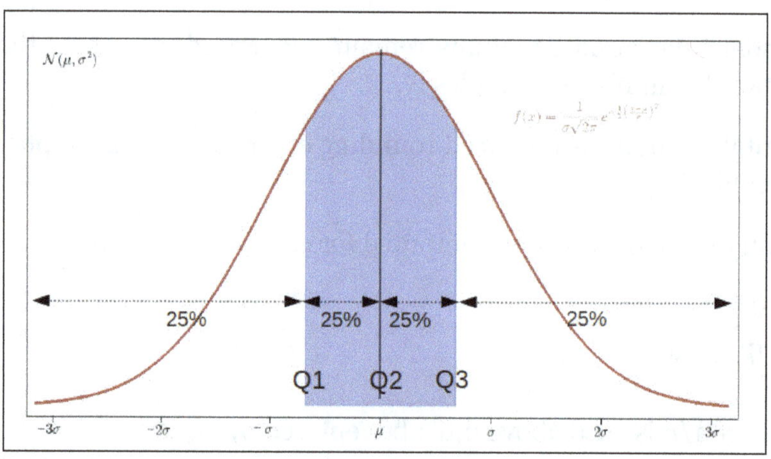

Probability density of a normal distribution, with quartiles shown. The area below the red curve is the same in the intervals $(-\infty,Q_1)$, (Q_1,Q_2), (Q_2,Q_3), and $(Q_3,+\infty)$.

In statistics and probability quantiles are cut points dividing the range of a probability distribution into continuous intervals with equal probabilities, or dividing the observations in a sample in the same way. There is one fewer quantile than the number of groups created. Thus quartiles are the three cut points that will divide a dataset into four equal-sized groups. Common quantiles have special names: for instance quartile, decile. The groups created are termed halves, thirds, quarters, etc., though sometimes the terms for the quantile are used for the groups created, rather than for the cut points.

q-quantiles are values that partition a finite set of values into q subsets of (nearly) equal sizes. There are $q - 1$ of the q-quantiles, one for each integer k satisfying $0 < k < q$. In some cases the value of a quantile may not be uniquely determined, as can be the case for the median (2-quantile) of a uniform probability distribution on a set of even size. Quantiles can also be applied to continuous distributions, providing a way to generalize rank statistics to continuous variables. When the cumulative distribution function of a random variable is known, the q-quantiles are the application of the quantile function (the inverse function of the cumulative distribution function) to the values $\{1/q, 2/q, \ldots, (q - 1)/q\}$.

Specialized Quantiles

Some q-quantiles have special names:

- The only 2-quantile is called the median.
- The 3-quantiles are called tertiles or terciles → T.
- The 4-quantiles are called quartiles → Q; the difference between upper and lower quartiles is also called the interquartile range, midspread or middle fifty → $IQR = Q_3 - Q_1$.
- The 5-quantiles are called quintiles → QU.
- The 6-quantiles are called sextiles → S.
- The 7-quantiles are called septiles.
- The 8-quantiles are called octiles.
- The 10-quantiles are called deciles → D.
- The 12-quantiles are called duo-deciles or dodeciles.
- The 16-quantiles are called hexadeciles → H.
- The 20-quantiles are called ventiles, vigintiles, or demi-deciles → V.
- The 100-quantiles are called percentiles → P.
- The 1000-quantiles have been called permilles or milliles, but these are rare and largely obsolete.

Quantiles of a Population

As in the computation of, for example, standard deviation, the estimation of a quantile depends upon whether one is operating with a statistical population or with a sample drawn from it. For a population, of discrete values or for a continuous population density, the k-th q-quantile is the data value where the cumulative distribution function crosses k/q. That is, x is a k-th q-quantile for a variable X if,

$$\Pr[X < x] \leq k/q \text{ or, equivalently, } \Pr[X \geq x] \geq 1 - k/q$$

and

$$\Pr[X \leq x] \geq k/q \text{ and } \Pr[X \geq x] \geq k/q.$$

For a finite population of N equally probable values indexed 1, ..., N from lowest to highest, the k-th q-quantile of this population can equivalently be computed via the value of $I_p = N\,k/q$. If I_p is not an integer, then round up to the next integer to get the appropriate index; the corresponding data value is the k-th q-quantile. On the other hand, if I_p is an integer then any number from the data value at that index to the data value of the next can be taken as the quantile, and it is conventional (though arbitrary) to take the average of those two values.

If, instead of using integers k and q, the "p-quantile" is based on a real number p with $0 < p < 1$ then p replaces k/q in the above formulas. Some software programs regard the minimum and maximum as the 0th and 100th percentile, respectively; however, such terminology is an extension beyond traditional statistics definitions.

The following two examples use the Nearest Rank definition of quantile with rounding. For an explanation of this definition.

Even-sized Population

Consider an ordered population of 10 data values {3, 6, 7, 8, 8, 10, 13, 15, 16, 20}. What are the 4-quantiles (the "quartiles") of this dataset?

Quartile	Calculation	Result
Zeroth quartile	Although not universally accepted, one can also speak of the zeroth quartile. This is the minimum value of the set, so the zeroth quartile in this example would be 3.	3
First quartile	The rank of the first quartile is 10×(1/4) = 2.5, which rounds up to 3, meaning that 3 is the rank in the population (from least to greatest values) at which approximately 1/4 of the values are less than the value of the first quartile. The third value in the population is 7.	7
Second quartile	The rank of the second quartile (same as the median) is 10×(2/4) = 5, which is an integer, while the number of values (10) is an even number, so the average of both the fifth and sixth values is taken—that is (8+10)/2 = 9, though any value from 8 through to 10 could be taken to be the median.	9

Third quartile	The rank of the third quartile is 10×(3/4) = 7.5, which rounds up to 8. The eighth value in the population is 15.	15
Fourth quartile	Although not universally accepted, one can also speak of the fourth quartile. This is the maximum value of the set, so the fourth quartile in this example would be 20. Under the Nearest Rank definition of quantile, the rank of the fourth quartile is the rank of the biggest number, so the rank of the fourth quartile would be 10.	20

So the first, second and third 4-quantiles (the "quartiles") of the dataset {3, 6, 7, 8, 8, 10, 13, 15, 16, 20} are {7, 9, 15}. If also required, the zeroth quartile is 3 and the fourth quartile is 20.

Odd-sized Population

Consider an ordered population of 11 data values {3, 6, 7, 8, 8, 9, 10, 13, 15, 16, 20}. What are the 4-quantiles (the "quartiles") of this dataset?

Quartile	Calculation	Result
Zeroth quartile	Although not universally accepted, one can also speak of the zeroth quartile. This is the minimum value of the set, so the zeroth quartile in this example would be 3.	3
First quartile	The first quartile is determined by 11×(1/4) = 2.75, which rounds up to 3, meaning that 3 is the rank in the population (from least to greatest values) at which approximately 1/4 of the values are less than the value of the first quartile. The third value in the population is 7.	7
Second quartile	The second quartile value (same as the median) is determined by 11×(2/4) = 5.5, which rounds up to 6. Therefore, 6 is the rank in the population (from least to greatest values) at which approximately 2/4 of the values are less than the value of the second quartile (or median). The sixth value in the population is 9.	9
Third quartile	The third quartile value for the original example above is determined by 11×(3/4) = 8.25, which rounds up to 9. The ninth value in the population is 15.	15
Fourth quartile	Although not universally accepted, one can also speak of the fourth quartile. This is the maximum value of the set, so the fourth quartile in this example would be 20. Under the Nearest Rank definition of quantile, the rank of the fourth quartile is the rank of the biggest number, so the rank of the fourth quartile would be 11.	20

So the first, second and third 4-quantiles (the "quartiles") of the dataset {3, 6, 7, 8, 8, 9, 10, 13, 15, 16, 20} are {7, 9, 15}. If also required, the zeroth quartile is 3 and the fourth quartile is 20.

Estimating Quantiles from a Sample

When one has a sample drawn from an unknown population, the cumulative distribution function and quantile function of the underlying population are not known and the task becomes that of estimating the quantiles. There are several methods. Mathematica, Matlab, R and GNU Octave programming languages include nine sample quantile methods. SAS includes five sample quantile methods, SciPy and Maple both include eight, EViews includes the six piecewise linear functions, Stata includes two, Python includes two, and Microsoft Excel includes two. Mathematica supports an arbitrary parameter for methods that allows for other, non-standard, methods.

In effect, the methods compute Q_p, the estimate for the k-th q-quantile, where $p = k/q$, from a sample of size N by computing a real valued index h. When h is an integer, the h-th smallest of the N values, x_h, is the quantile estimate. Otherwise a rounding or interpolation scheme is used to compute the quantile estimate from h, $x_{\lfloor h \rfloor}$, and $x_{\lceil h \rceil}$.

The estimate types and interpolation schemes used include:

Type	h	Q_p	Notes
R-1, SAS-3, Maple-1	Np + 1/2	$x_{\lceil h - 1/2 \rceil}$	Inverse of empirical distribution function.
R-2, SAS-5, Maple-2, Stata	Np + 1/2	$(x_{\lceil h - 1/2 \rceil} + x_{\lfloor h + 1/2 \rfloor}) / 2$	The same as R-1, but with averaging at discontinuities.
R-3, SAS-2	Np	$x_{\lfloor h \rceil}$	The observation numbered closest to Np. Here, ⌊h⌉ indicates rounding to the nearest integer, choosing the even integer in the case of a tie.
R-4, SAS-1, SciPy-(0,1), Maple-3	Np	$x_{\lfloor h \rfloor} + (h - \lfloor h \rfloor)(x_{\lfloor h \rfloor + 1} - x_{\lfloor h \rfloor})$	Linear interpolation of the empirical distribution function.
R-5, SciPy-(.5,.5), Maple-4	Np + 1/2	$x_{\lfloor h \rfloor} + (h - \lfloor h \rfloor)(x_{\lfloor h \rfloor + 1} - x_{\lfloor h \rfloor})$	Piecewise linear function where the knots are the values midway through the steps of the empirical distribution function.
R-6, Excel, Python, SAS-4, SciPy-(0,0), Maple-5, Stata-altdef	(N + 1)p	$x_{\lfloor h \rfloor} + (h - \lfloor h \rfloor)(x_{\lfloor h \rfloor + 1} - x_{\lfloor h \rfloor})$	Linear interpolation of the expectations for the order statistics for the uniform distribution on [0,1]. That is, it is the linear interpolation between points (p_h, x_h), where $p_h = h/(N+1)$ is the probability that the last of (N+1) randomly drawn values will not exceed the h-th smallest of the first N randomly drawn values.
R-7, Excel, Python, SciPy-(1,1), Maple-6, NumPy, Julia	(N − 1)p + 1	$x_{\lfloor h \rfloor} + (h - \lfloor h \rfloor)(x_{\lfloor h \rfloor + 1} - x_{\lfloor h \rfloor})$	Linear interpolation of the modes for the order statistics for the uniform distribution on [0,1].
R-8, SciPy-(1/3,1/3), Maple-7	(N + 1/3)p + 1/3	$x_{\lfloor h \rfloor} + (h - \lfloor h \rfloor)(x_{\lfloor h \rfloor + 1} - x_{\lfloor h \rfloor})$	Linear interpolation of the approximate medians for order statistics.
R-9, SciPy-(3/8,3/8), Maple-8	(N + 1/4)p + 3/8	$x_{\lfloor h \rfloor} + (h - \lfloor h \rfloor)(x_{\lfloor h \rfloor + 1} - x_{\lfloor h \rfloor})$	The resulting quantile estimates are approximately unbiased for the expected order statistics if x is normally distributed.

- R-1 through R-3 are piecewise constant, with discontinuities.
- R-4 and following are piecewise linear, without discontinuities, but differ in how h is computed.
- R-3 and R-4 are not symmetric in that they do not give $h = (N + 1)/2$ when $p = 1/2$.
- Excel's PERCENTILE.EXC and Python's default "exclusive" method are equivalent to R-6.
- Excel's PERCENTILE and PERCENTILE.INC and Python's optional "inclusive" method are equivalent to R-7.
- Packages differ in how they estimate quantiles beyond the lowest and highest values in the sample. Choices include returning an error value, computing linear extrapolation, or assuming a constant value.

The standard error of a quantile estimate can in general be estimated via the bootstrap. The Maritz–Jarrett method can also be used.

QUARTILE

A quartile is a type of quantile which divides the number of data points into four more or less equal parts, or quarters. The first quartile (Q_1) is defined as the middle number between the smallest number and the median of the data set. It is also known as the lower quartile or the 25th empirical quartile and it marks where 25% of the data is below or to the left of it (if data is ordered on a timeline from smallest to largest). The second quartile (Q_2) is the median of the data and 50% of the data lies below this point. The third quartile (Q_3) is the middle value between the median and the highest value of the data set. It is also known as the upper quartile or the 75th empirical quartile and 75% of the data lies below this point. Due to the fact that the data needs to be ordered from smallest to largest in order to compute quartiles, quartiles are a form of Order statistic.

Along with the minimum and the maximum of the data, which are also quartiles, the three quartiles described above provide a five-number summary of the data. This summary is important in statistics because it provides information about both the center and the spread of the data. Knowing the lower and upper quartile provides information on how big the spread is and if the dataset is skewed toward one side. Since quartiles divide the number of data points evenly, the range is not the same between quartiles (ie. Q_3-Q_2 ≠ Q_2-Q_1). While the maximum and minimum also show the spread of the data, the upper and lower quartiles can provide more detailed information on the location of specific data points, the presence of outliers in the data, and the difference in spread between the middle 50% of the data and the outer data points.

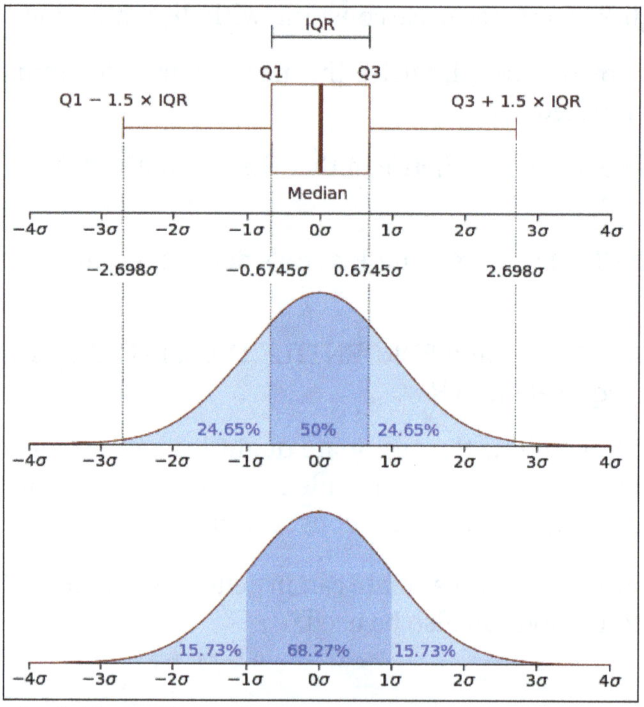

Boxplot (with quartiles and an interquartile range) and a probability density function (pdf) of a normal N(0,1σ²) population.

Symbol	Names	Definition
Q_1	First quartile lower quartile 25th percentile.	Splits off the lowest 25% of data from the highest 75%.
Q_2	Second quartile median 50th percentile.	Cuts data set in half.
Q_3	Third quartile upper quartile 75th percentile.	Splits off the highest 25% of data from the lowest 75%.

Computing Methods

Discrete Distributions

For discrete distributions, there is no universal agreement on selecting the quartile values.

Method 1

- Use the median to divide the ordered data set into two halves:
  - If there is an odd number of data points in the original ordered data set, do not include the median (the central value in the ordered list) in either half.

   - If there is an even number of data points in the original ordered data set, split this data set exactly in half.
- The lower quartile value is the median of the lower half of the data. The upper quartile value is the median of the upper half of the data.

This rule is employed by the TI-83 calculator boxplot and "1-Var Stats" functions.

Method 2

- Use the median to divide the ordered data set into two halves:
   - If there are an odd number of data points in the original ordered data set, include the median (the central value in the ordered list) in both halves.
   - If there are an even number of data points in the original ordered data set, split this data set exactly in half.
- The lower quartile value is the median of the lower half of the data. The upper quartile value is the median of the upper half of the data.

The values found by this method are also known as "Tukey's hinges".

Method 3

- If there are even numbers of data points, then Method 3 is the same as either method above.
- If there are ($4n$+1) data points, then the lower quartile is 25% of the nth data value plus 75% of the (n+1)th data value; the upper quartile is 75% of the ($3n$+1) th data point plus 25% of the ($3n$+2)th data point.
- If there are ($4n$+3) data points, then the lower quartile is 75% of the (n+1)th data value plus 25% of the (n+2)th data value; the upper quartile is 25% of the ($3n$+2)th data point plus 75% of the ($3n$+3)th data point.

Method 4

If we have an ordered dataset $x_1, x_2, \ldots, x_n$, we can interpolate between data points to find the p th empirical quantile if x_i is in the $i/(n+1)$ quantile. If we denote the integer part of a number by a by $[a]$, then the empirical quantile function is given by,

$$q(p) = x_{(k)} + \alpha(x_{(k+1)} - x_{(k)}),$$

where $k = [p(n+1)]$ and $\alpha = p(n+1) - [p(n+1)]$

To find the first, second, and third quartiles of the dataset we would evaluate $q(0.25)$, $q(0.5)$, and $q(0.75)$ respectively.

Example: Ordered Data Set: 6, 7, 15, 36, 39, 40, 41, 42, 43, 47, 49

	Method 1	Method 2	Method 3	Method 4
Q_1	15	25.5	20.25	15
Q_2	40	40	40	40
Q_3	43	42.5	42.75	43

Example: Ordered Data Set: 7, 15, 36, 39, 40, 41

As there are an even number of data points, all three methods give the same results.

	Method 1	Method 2	Method 3	Method 4
Q_1	15	15	15	13
Q_2	37.5	37.5	37.5	37.5
Q_3	40	40	40	40.25

Continuous Probability Distributions

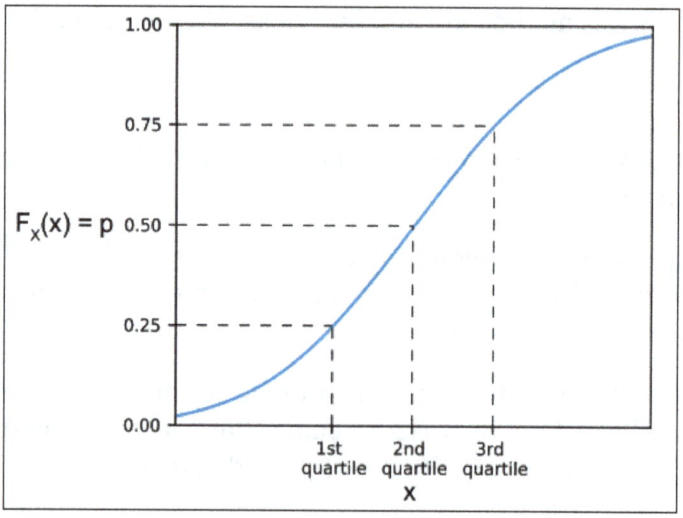

Quartiles on a cumulative distribution function of a normal distribution.

If we define a continuous probability distributions as $P(X)$ where X is a real valued random variable, its cumulative distribution function (CDF) is given by,

$$F_X(x) = P(X \leq x)$$

The CDF gives the probability that the random variable X is less than the value x. Therefore, the first quartile is the value of x when $F_X(x) = 0.25$, the second quartile is x when $F_X(x) = 0.5$, and the third quartile is x when $F_X(x) = 0.75$. The values of $Q(p)$ can be found with the quantile function $Q(p)$ where $p = 0.25$ for the first quartile, $p = 0.5$ for the second quartile, and $p = 0.75$ for the third quartile. The quantile function is the inverse of the cumulative distribution function if the cumulative distribution function is monotonically increasing.

Outliers

There are methods by which to check for outliers in the discipline of statistics and statistical analysis. Outliers could be a result from a shift in the location (mean) or in the scale (variability) of the process of interest. Outliers could also may be evidence of a sample population that has a non-normal distribution or of a contaminated population data set. Consequently, as is the basic idea of descriptive statistics, when encountering an outlier, we have to explain this value by further analysis of the cause or origin of the outlier. In cases of extreme observations, which are not an infrequent occurrence, the typical values must be analyzed. In the case of quartiles, the Interquartile Range (IQR) may be used to characterize the data when there may be extremities that skew the data; the interquartile range is a relatively robust statistic (also sometimes called "resistance") compared to the range and standard deviation. There is also a mathematical method to check for outliers and determining "fences", upper and lower limits from which to check for outliers.

After determining the first and third quartiles and the interquartile range as outlined above, then fences are calculated using the following formula:

$$\text{Lower fence} = Q_1 - 1.5(\text{IQR})$$

$$\text{Upper fence} = Q_3 + 1.5(\text{IQR}),$$

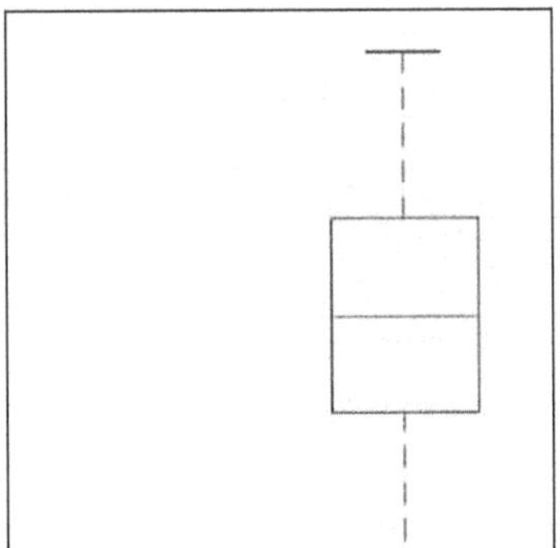

Boxplot Diagram with Outliers.

where Q_1 and Q_3 are the first and third quartiles, respectively. The lower fence is the "lower limit" and the upper fence is the "upper limit" of data, and any data lying outside these defined bounds can be considered an outlier. Anything below the Lower fence or above the Upper fence can be considered such a case. The fences provide a guideline by which to define an outlier, which may be defined in other ways. The fences define a "range" outside of which an outlier exists; a way to picture this is a boundary of a fence, outside of which are "outsiders" as opposed to outliers. It is common for the lower and upper fences along with the outliers to be represented by a boxplot. For a boxplot, only the vertical heights correspond to the visualized data set while horizontal width of the box is irrelevant. Outliers located outside the fences in a boxplot can be marked as any choice of symbol, such as an "x" or "o". The fences are sometimes also referred to as "whiskers" while the entire plot visual is called a "box-and-whisker" plot.

When spotting an outlier in the data set by calculating the interquartile ranges and box-plot features, it might be simple to mistakenly view it as evidence that the population is non-normal or that the sample is contaminated. However, this method should not take place of a hypothesis test for determining normality of the population. The significance of the outliers vary depending on the sample size. If the sample is small, then it is more probable to get interquartile ranges that are unrepresentatively small, leading to narrower fences. Therefore, it would be more likely to find data that are marked as outliers.

Computer Software for Quartiles

Excel:

The Excel function *QUARTILE(array, quart)* provides the desired quartile value for a given array of data. In the *Quartile* function, array is the dataset of numbers that is being analyzed and quart is any of the following 5 values depending on which quartile is being calculated.

Quart	Output QUARTILE Value
0	Minimum value
1	Lower Quartile (25th percentile)
2	Median
3	Upper Quartile (75th percentile)
4	Maximum value

MATLAB:

In order to calculate quartiles in Matlab, the function *quantile(A,p)* can be used. Where A is the vector of data being analyzed and p is the percentage that relates to the quartiles as stated below.

p	Output QUARTILE Value
0	Minimum value
0.25	Lower Quartile (25th percentile)
0.5	Median
0.75	Upper Quartile (75th percentile)
1	Maximum value

QUARTILE DEVIATION AND ITS COEFFICIENT

The Quartile Deviation is a simple way to estimate the spread of a distribution about a measure of its central tendency (usually the mean). So, it gives you an idea about the range within which the central 50% of your sample data lies. Consequently, based on the quartile deviation, the Coefficient of Quartile Deviation can be defined, which makes it easy to compare the spread of two or more different distributions.

The Quartile Deviation

Formally, the Quartile Deviation is equal to the half of the Inter-Quartile Range and thus we can write it as,

$$Q_d = \frac{Q_3 - Q_1}{2}$$

Therefore, we also call it the Semi Inter-Quartile Range.

- The Quartile Deviation doesn't take into account the extreme points of the distribution. Thus, the dispersion or the spread of only the central 50% data is considered.
- If the scale of the data is changed, the Qd also changes in the same ratio.
- It is the best measure of dispersion for open-ended systems (which have open-ended extreme ranges).
- Also, it is less affected by sampling fluctuations in the dataset as compared to the range (another measure of dispersion).
- Since it is solely dependent on the central values in the distribution, if in any experiment, these values are abnormal or inaccurate, the result would be affected drastically.

The Coefficient of Quartile Deviation

Based on the quartiles, a relative measure of dispersion, known as the Coefficient of Quartile Deviation, can be defined for any distribution. It is formally defined as,

$$\text{Coefficient of Quartile Deviation} = \frac{Q_3 - Q_1}{Q_3 + Q_1} \times 100$$

Since it involves a ratio of two quantities of the same dimensions, it is unit-less. Thus, it can act as a suitable parameter for comparing two or more different datasets which may or may not involve quantities with the same dimensions.

VARIANCE

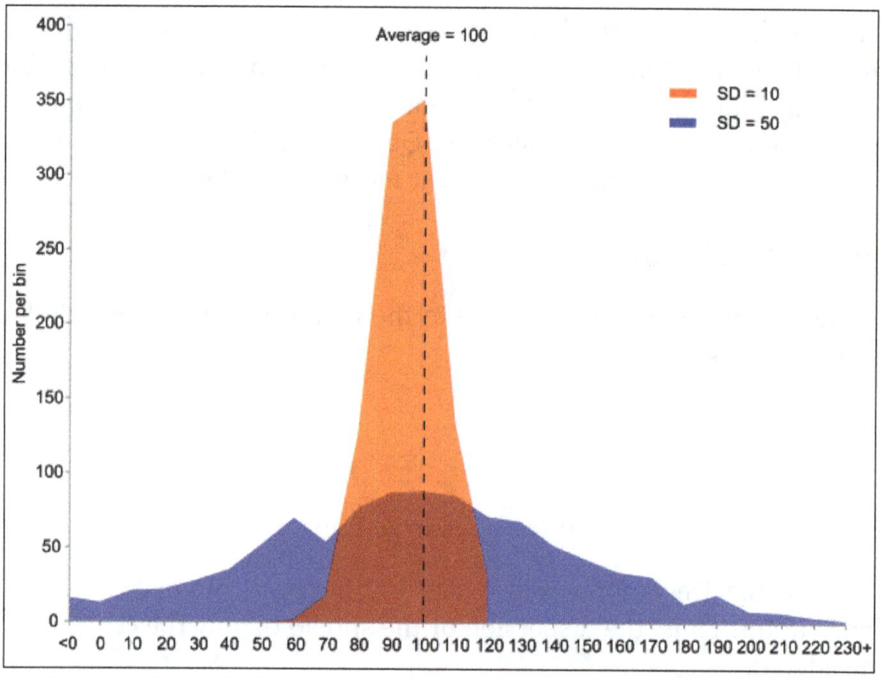

Example of samples from two populations with the same mean but different variances. The red population has mean 100 and variance 100 (SD=10) while the blue population has mean 100 and variance 2500 (SD=50).

In probability theory and statistics, variance is the expectation of the squared deviation of a random variable from its mean. Informally, it measures how far a set of (random) numbers are spread out from their average value. Variance has a central role in statistics, where some ideas that use it include descriptive statistics, statistical inference, hypothesis testing, goodness of fit, and Monte Carlo sampling. Variance is an important tool in the sciences, where statistical analysis of data is common. The variance is the square of the standard deviation, the second central moment of a distribution, and the covariance of the random variable with itself, and it is often represented by σ^2, s^2 or Var(X).

The variance of a random variable X is the expected value of the squared deviation from the mean of X, $\mu = \mathrm{E}[X]$:

$$\mathrm{Var}(X) = \mathrm{E}\left[(X-\mu)^2\right]$$

This definition encompasses random variables that are generated by processes that are discrete, continuous, neither, or mixed. The variance can also be thought of as the covariance of a random variable with itself:

$$\mathrm{Var}(X) = \mathrm{Cov}(X, X).$$

The variance is also equivalent to the second cumulant of a probability distribution that generates X. The variance is typically designated as $\operatorname{Var}(X)$, σ_X^2, or simply σ^2. The expression for the variance can be expanded:

$$\begin{aligned}\operatorname{Var}(X) &= \operatorname{E}\left[(X-\operatorname{E}[X])^2\right] \\ &= \operatorname{E}\left[X^2 - 2X\operatorname{E}[X] + \operatorname{E}[X]^2\right] \\ &= \operatorname{E}\left[X^2\right] - 2\operatorname{E}[X]\operatorname{E}[X] + \operatorname{E}[X]^2 \\ &= \operatorname{E}\left[X^2\right] - \operatorname{E}[X]^2\end{aligned}$$

In other words, the variance of *X* is equal to the mean of the square of *X* minus the square of the mean of *X*. This equation should not be used for computations using floating point arithmetic because it suffers from catastrophic cancellation if the two components of the equation are similar in magnitude. There exist numerically stable alternatives.

Discrete Random Variable

If the generator of random variable X is discrete with probability mass function $x_1 \mapsto p_1, x_2 \mapsto p_2, \ldots, x_n \mapsto p_n$ then,

$$\operatorname{Var}(X) = \sum_{i=1}^{n} p_i \cdot (x_i - \mu)^2,$$

or equivalently,

$$\operatorname{Var}(X) = \left(\sum_{i=1}^{n} p_i x_i^2\right) - \mu^2,$$

where μ is the expected value, i.e.,

$$\mu = \sum_{i=1}^{n} p_i x_i.$$

(When such a discrete weighted variance is specified by weights whose sum is not 1, then one divides by the sum of the weights.)

The variance of a set of n equally likely values can be written as,

$$\operatorname{Var}(X) = \frac{1}{n}\sum_{i=1}^{n} (x_i - \mu)^2,$$

where μ is the average value, i.e.,

$$\mu = \frac{1}{n}\sum_{i=1}^{n} x_i.$$

The variance of a set of n equally likely values can be equivalently expressed, without directly referring to the mean, in terms of squared deviations of all points from each other:

$$\operatorname{Var}(X) = \frac{1}{n^2}\sum_{i=1}^{n}\sum_{j=1}^{n}\frac{1}{2}(x_i - x_j)^2 = \frac{1}{n^2}\sum_{i}\sum_{j>i}(x_i - x_j)^2.$$

Absolutely Continuous Random Variable

If the random variable X has a probability density function $f(x)$, and $F(x)$ is the corresponding cumulative distribution function, then:

$$\begin{aligned}
\operatorname{Var}(X) = \sigma^2 &= \int_{\mathbb{R}} (x-\mu)^2 f(x)dx \\
&= \int_{\mathbb{R}} x^2 f(x)dx - 2\mu\int_{\mathbb{R}} xf(x)dx + \int_{\mathbb{R}} \mu^2 f(x)dx \\
&= \int_{\mathbb{R}} x^2 dF(x) - 2\mu\int_{\mathbb{R}} xdF(x) + \mu^2\int_{\mathbb{R}} dF(x) \\
&= \int_{\mathbb{R}} x^2 dF(x) - 2\mu\cdot\mu + \mu^2\cdot 1 \\
&= \int_{\mathbb{R}} x^2 dF(x) - \mu^2,
\end{aligned}$$

or equivalently,

$$\operatorname{Var}(X) = \int_{\mathbb{R}} x^2 f(x)dx - \mu^2,$$

where μ is the expected value of X given by,

$$\mu = \int_{\mathbb{R}} xf(x)dx = \int_{\mathbb{R}} xdF(x).$$

In these formulas, the integrals with respect to dx and $dF(x)$ are Lebesgue and Lebesgue–Stieltjes integrals, respectively.

If the function $x^2 f(x)$ is Riemann-integrable on every finite interval $[a,b]\subset\mathbb{R}$ then

$$\operatorname{Var}(X) = \int_{-\infty}^{+\infty} x^2 f(x)dx - \mu^2,$$

where the integral is an improper Riemann integral.

Examples:

Exponential Distribution

The exponential distribution with parameter λ is a continuous distribution whose probability density function is given by,

$$f(x) = \lambda e^{-\lambda x}$$

on the interval [0, ∞). Its mean can be shown to be,

$$\mathrm{E}[X] = \int_0^\infty \lambda x e^{-\lambda x}\,dx = \frac{1}{\lambda}.$$

Using integration by parts and making use of the expected value already calculated:

$$\begin{aligned}\mathrm{E}[X^2] &= \int_0^\infty \lambda x^2 e^{-\lambda x}\,dx \\ &= \left[-x^2 e^{-\lambda x}\right]_0^\infty + \int_0^\infty 2x e^{-\lambda x}\,dx \\ &= 0 + \frac{2}{\lambda}\mathrm{E}[X] \\ &= \frac{2}{\lambda^2}.\end{aligned}$$

Thus, the variance of X is given by,

$$\mathrm{Var}(X) = \mathrm{E}[X^2] - \mathrm{E}[X]^2 = \frac{2}{\lambda^2} - \left(\frac{1}{\lambda}\right)^2 = \frac{1}{\lambda^2}.$$

Fair Die

A fair six-sided die can be modeled as a discrete random variable, X, with outcomes 1 through 6, each with equal probability 1/6. The expected value of X is $(1+2+3+4+5+6)/6 = 7/2$. Therefore, the variance of X is,

$$\begin{aligned}\mathrm{Var}(X) &= \sum_{i=1}^{6} \frac{1}{6}\left(i - \frac{7}{2}\right)^2 \\ &= \frac{1}{6}\left((-5/2)^2 + (-3/2)^2 + (-1/2)^2 + (1/2)^2 + (3/2)^2 + (5/2)^2\right) \\ &= \frac{35}{12} \approx 2.92.\end{aligned}$$

The general formula for the variance of the outcome, X, of an n-sided die is,

$$\begin{aligned}\operatorname{Var}(X) &= \operatorname{E}(X^2) - (\operatorname{E}(X))^2 \\ &= \frac{1}{n}\sum_{i=1}^{n} i^2 - \left(\frac{1}{n}\sum_{i=1}^{n} i\right)^2 \\ &= \frac{(n+1)(2n+1)}{6} - \left(\frac{n+1}{2}\right)^2 \\ &= \frac{n^2-1}{12}.\end{aligned}$$

Commonly used Probability Distributions

The following table lists the variance for some commonly used probability distributions.

Name of the probability distribution	Probability distribution function	Variance
Binomial distribution	$\Pr(X=k)=\binom{n}{k}p^k(1-p)^{n-k}$	$np(1-p)$
Geometric distribution	$\Pr(X=k)=(1-p)^{k-1}p$	$\frac{(1-p)}{p^2}$
Normal distribution	$f(x\mid\mu,\sigma^2)=\frac{1}{\sqrt{2\pi\sigma^2}}e^{-\frac{(x-\mu)^2}{2\sigma^2}}$	σ^2
Uniform distribution (continuous)	$f(x\mid a,b)=\begin{cases}\frac{1}{b-a} & \text{for } a\le x\le b,\\ 0 & \text{for } x<a \text{ or } x>b\end{cases}$	$\frac{(b-a)^2}{12}$
Exponential distribution	$f(x\mid\lambda)=\lambda e^{-\lambda x}$	$\frac{1}{\lambda^2}$

Properties

Variance is non-negative because the squares are positive or zero:

$$\operatorname{Var}(X)\ge 0.$$

The variance of a constant is zero,

$$\operatorname{Var}(a)=0.$$

If the variance of a random variable is 0, then it is a constant. That is, it always has the same value:

$$\mathrm{Var}(X) = 0 \Leftrightarrow P(X = a) = 1.$$

Variance is invariant with respect to changes in a location parameter. That is, if a constant is added to all values of the variable, the variance is unchanged:

$$\mathrm{Var}(X + a) = \mathrm{Var}(X).$$

If all values are scaled by a constant, the variance is scaled by the square of that constant:

$$\mathrm{Var}(aX) = a^2 \, \mathrm{Var}(X).$$

The variance of a sum of two random variables is given by,

$$\mathrm{Var}(aX + bY) = a^2 \, \mathrm{Var}(X) + b^2 \, \mathrm{Var}(Y) + 2ab \, \mathrm{Cov}(X,Y),$$

$$\mathrm{Var}(aX - bY) = a^2 \, \mathrm{Var}(X) + b^2 \, \mathrm{Var}(Y) - 2ab \, \mathrm{Cov}(X,Y),$$

where $\mathrm{Cov}(\cdot, \cdot)$ is the covariance. In general we have for the sum of N random variables $\{X_1, \ldots, X_N\}$:

$$\mathrm{Var}\left(\sum_{i=1}^{N} X_i\right) = \sum_{i,j=1}^{N} \mathrm{Cov}(X_i, X_j)$$

$$= \sum_{i=1}^{N} \mathrm{Var}(X_i) + \sum_{i \neq j} \mathrm{Cov}(X_i, X_j).$$

These results lead to the variance of a linear combination as:

$$\mathrm{Var}\left(\sum_{i=1}^{N} a_i X_i\right) = \sum_{i,j=1}^{N} a_i a_j \, \mathrm{Cov}(X_i, X_j)$$

$$= \sum_{i=1}^{N} a_i^2 \, \mathrm{Var}(X_i) + \sum_{i \neq j} a_i a_j \, \mathrm{Cov}(X_i, X_j)$$

$$= \sum_{i=1}^{N} a_i^2 \, \mathrm{Var}(X_i) + 2 \sum_{1 \le i < j \le N} a_i a_j \, \mathrm{Cov}(X_i, X_j).$$

If the random variables $X_1, \ldots, X_N$ are such that,

$$\mathrm{Cov}(X_i, X_j) = 0, \; \forall \, (i \neq j),$$

they are said to be uncorrelated. It follows immediately from the expression given earlier that if the random variables $X_1,\ldots,X_N$ are uncorrelated, then the variance of their sum is equal to the sum of their variances, or, expressed symbolically:

$$\operatorname{Var}\left(\sum_{i=1}^{N} X_i\right) = \sum_{i=1}^{N} \operatorname{Var}(X_i).$$

Since independent random variables are always uncorrelated, the equation above holds in particular when the random variables $X_1,\ldots,X_n$ are independent. Thus independence is sufficient but not necessary for the variance of the sum to equal the sum of the variances.

Issues of Finiteness

If a distribution does not have a finite expected value, as is the case for the Cauchy distribution, then the variance cannot be finite either. However, some distributions may not have a finite variance despite their expected value being finite. An example is a Pareto distribution whose index k satisfies $1 < k \leq 2$.

Sum of Uncorrelated Variables (Bienaymé Formula)

One reason for the use of the variance in preference to other measures of dispersion is that the variance of the sum (or the difference) of uncorrelated random variables is the sum of their variances:

$$\operatorname{Var}\left(\sum_{i=1}^{n} X_i\right) = \sum_{i=1}^{n} \operatorname{Var}(X_i).$$

This statement is called the Bienaymé formula and was discovered in 1853. It is often made with the stronger condition that the variables are independent, but being uncorrelated suffices. So if all the variables have the same variance σ^2, then, since division by n is a linear transformation, this formula immediately implies that the variance of their mean is,

$$\operatorname{Var}\left(\bar{X}\right) = \operatorname{Var}\left(\frac{1}{n}\sum_{i=1}^{n} X_i\right) = \frac{1}{n^2}\sum_{i=1}^{n} \operatorname{Var}\left(X_i\right) = \frac{1}{n^2} n\sigma^2 = \frac{\sigma^2}{n}.$$

That is, the variance of the mean decreases when n increases. This formula for the variance of the mean is used in the definition of the standard error of the sample mean, which is used in the central limit theorem.

To prove the initial statement, it suffices to show that,

$$\operatorname{Var}(X+Y) = \operatorname{Var}(X) + \operatorname{Var}(Y).$$

The general result then follows by induction. Starting with the definition,

$$\begin{aligned}\operatorname{Var}(X+Y) &= \mathrm{E}[(X+Y)^2]-(\mathrm{E}[X+Y])^2 \\ &= \mathrm{E}[X^2+2XY+Y^2]-(\mathrm{E}[X]+\mathrm{E}[Y])^2.\end{aligned}$$

Using the linearity of the expectation operator and the assumption of independence (or uncorrelatedness) of X and Y, this further simplifies as follows:

$$\begin{aligned}\operatorname{Var}(X+Y) &= \mathrm{E}[X^2]+2\mathrm{E}[XY]+\mathrm{E}[Y^2]-(\mathrm{E}[X]^2+2\mathrm{E}[X]\mathrm{E}[Y]+\mathrm{E}[Y]^2) \\ &= \mathrm{E}[X^2]+\mathrm{E}[Y^2]-\mathrm{E}[X]^2-\mathrm{E}[Y]^2 \\ &= \operatorname{Var}(X)+\operatorname{Var}(Y).\end{aligned}$$

Sum of Correlated Variables

With Correlation and Fixed Sample Size

In general the variance of the sum of n variables is the sum of their covariances:

$$\operatorname{Var}\left(\sum_{i=1}^{n} X_i\right)=\sum_{i=1}^{n}\sum_{j=1}^{n}\operatorname{Cov}(X_i,X_j)=\sum_{i=1}^{n}\operatorname{Var}(X_i)+2\sum_{1\le i<j\le n}\operatorname{Cov}(X_i,X_j).$$

(The second equality comes from the fact that $\operatorname{Cov}(X_i,X_i)=\operatorname{Var}(X_i)$.)

Here $\operatorname{Cov}(\cdot,\cdot)$ is the covariance, which is zero for independent random variables (if it exists). The formula states that the variance of a sum is equal to the sum of all elements in the covariance matrix of the components. The next expression states equivalently that the variance of the sum is the sum of the diagonal of covariance matrix plus two times the sum of its upper triangular elements (or its lower triangular elements); this emphasizes that the covariance matrix is symmetric. This formula is used in the theory of Cronbach's alpha in classical test theory.

So if the variables have equal variance σ^2 and the average correlation of distinct variables is ρ, then the variance of their mean is,

$$Var(\bar{X})=\frac{\sigma^2}{n}+\frac{n-1}{n}\rho\sigma^2.$$

This implies that the variance of the mean increases with the average of the correlations. In other words, additional correlated observations are not as effective as additional independent observations at reducing the uncertainty of the mean. Moreover, if the variables have unit variance, for example if they are standardized, then this simplifies to:

$$\operatorname{Var}(\bar{X})=\frac{1}{n}+\frac{n-1}{n}\rho.$$

This formula is used in the Spearman–Brown prediction formula of classical test theory. This converges to ρ if n goes to infinity, provided that the average correlation remains constant or converges too. So for the variance of the mean of standardized variables with equal correlations or converging average correlation we have,

$$\lim_{n\to\infty} \operatorname{Var}(\overline{X}) = \rho.$$

Therefore, the variance of the mean of a large number of standardized variables is approximately equal to their average correlation. This makes clear that the sample mean of correlated variables does not generally converge to the population mean, even though the law of large numbers states that the sample mean will converge for independent variables.

I.i.d. with Random Sample size

There are cases when a sample is taken without knowing, in advance, how many observations will be acceptable according to some criterion. In such cases, the sample size N is a random variable whose variation adds to the variation of X, such that,

$$\operatorname{Var}(\sum X) = E(N)\operatorname{Var}(X) + \operatorname{Var}(N)E^2(X).$$

If N has a Poisson distribution, then E(N) = Var(N) with estimator N=n. So, the estimator of Var(∑X) becomes $nS_X^2 + nXbar^2$ giving,

$$\text{standard error(Xbar)} = \sqrt{[(S_X^2 + Xbar^2)/n]}.$$

Matrix Notation for the Variance of a Linear Combination

Define X as a column vector of n random variables $X_1,\ldots,X_n$, and c as a column vector of n scalars $c_1,\ldots,c_n$. Therefore, $c^T X$ is a linear combination of these random variables, where c^T denotes the transpose of c. Also let Σ be the covariance matrix of X. The variance of $c^T X$ s then given by:

$$\operatorname{Var}(c^T X) = c^T \Sigma c.$$

This implies that the variance of the mean can be written as (with a column vector of ones),

$$\operatorname{Var}(\overline{x}) = \operatorname{Var}(1/n \cdot 1'X) = 1/n^2 \cdot 1'\Sigma 1.$$

Weighted Sum of Variables

The scaling property and the Bienaymé formula, along with the property of the covariance Cov(aX, bY) = ab Cov(X, Y) jointly imply that,

$$\operatorname{Var}(aX \pm bY) = a^2 \operatorname{Var}(X) + b^2 \operatorname{Var}(Y) \pm 2ab \operatorname{Cov}(X,Y).$$

This implies that in a weighted sum of variables, the variable with the largest weight will have a disproportionally large weight in the variance of the total. For example, if X and Y are uncorrelated and the weight of X is two times the weight of Y, then the weight of the variance of X will be four times the weight of the variance of Y.

The expression above can be extended to a weighted sum of multiple variables:

$$\operatorname{Var}\left(\sum_i^n a_i X_i\right) = \sum_{i=1}^n a_i^2 \operatorname{Var}(X_i) + 2\sum\sum_{1 \le i < j \le n} a_i a_j \operatorname{Cov}(X_i, X_j)$$

Product of Independent Variables

If two variables X and Y are independent, the variance of their product is given by,

$$\operatorname{Var}(XY) = [\operatorname{E}(X)]^2 \operatorname{Var}(Y) + [\operatorname{E}(Y)]^2 \operatorname{Var}(X) + \operatorname{Var}(X)\operatorname{Var}(Y).$$

Equivalently, using the basic properties of expectation, it is given by,

$$\operatorname{Var}(XY) = \operatorname{E}(X^2)\operatorname{E}(Y^2) - [\operatorname{E}(X)]^2[\operatorname{E}(Y)]^2.$$

Product of Statistically Dependent Variables

In general, if two variables are statistically dependent, the variance of their product is given by:

$$\begin{aligned}\operatorname{Var}(XY) &= \operatorname{E}[X^2Y^2] - [\operatorname{E}(XY)]^2 \\ &= \operatorname{Cov}(X^2, Y^2) + \operatorname{E}(X^2)\operatorname{E}(Y^2) - [\operatorname{E}(XY)]^2 \\ &= \operatorname{Cov}(X^2, Y^2) + (\operatorname{Var}(X) + [\operatorname{E}(X)]^2)(\operatorname{Var}(Y) + [\operatorname{E}(Y)]^2) \\ &\quad - [\operatorname{Cov}(X, Y) + \operatorname{E}(X)\operatorname{E}(Y)]^2\end{aligned}$$

Decomposition

The general formula for variance decomposition or the law of total variance is: If X and Y are two random variables, and the variance of X exists, then:

$$\operatorname{Var}[X] = \operatorname{E}(\operatorname{Var}[X \mid Y]) + \operatorname{Var}(\operatorname{E}[X \mid Y]).$$

The conditional expectation $\operatorname{E}(X \mid Y)$ of X given Y, and the conditional variance $Var(X \mid Y)$ may be understood as follows. Given any particular value y of the random variable Y, there is a conditional expectation $\operatorname{E}(X \mid Y = y)$ given the event $Y = y$. This quantity depends on the particular value y; it is a function $g(y) = \operatorname{E}(X \mid Y = y)$. That same function evaluated at the random variable Y is the conditional expectation,

$$\operatorname{E}(X \mid Y) = g(Y).$$

In particular, if Y is a discrete random variable assuming possible values $y_1, y_2, y_3 \ldots$ with corresponding probabilities $p_1, p_2, p_3 \ldots$, then in the formula for total variance, the first term on the right-hand side becomes,

$$\mathrm{E}(\mathrm{Var}[X \mid Y]) = \sum_i p_i \sigma_i^2,$$

where $\sigma_i^2 = Var[X \mid Y = y_i]$. Similarly, the second term on the right-hand side becomes,

$$\mathrm{Var}(\mathrm{E}[X \mid Y]) = \sum_i p_i \mu_i^2 - \left(\sum_i p_i \mu_i\right)^2 = \sum_i p_i \mu_i^2 - \mu^2,$$

where $\mu_i = \mathrm{E}[X \mid Y = y_i]$ and $\mu = \sum_i p_i \mu_i$.. Thus the total variance is given by,

$$\mathrm{Var}[X] = \sum_i p_i \sigma_i^2 + \left(\sum_i p_i \mu_i^2 - \mu^2\right).$$

A similar formula is applied in analysis of variance, where the corresponding formula is,

$$MS_{\text{total}} = MS_{\text{between}} + MS_{\text{within}};$$

here MS refers to the Mean of the Squares. In linear regression analysis the corresponding formula is,

$$MS_{\text{total}} = MS_{\text{regression}} + MS_{\text{residual}}.$$

This can also be derived from the additivity of variances, since the total (observed) score is the sum of the predicted score and the error score, where the latter two are uncorrelated.

Similar decompositions are possible for the sum of squared deviations (sum of squares, SS):

$$SS_{\text{total}} = SS_{\text{between}} + SS_{\text{within}},$$

$$SS_{\text{total}} = SS_{\text{regression}} + SS_{\text{residual}}.$$

Calculation from the CDF

The population variance for a non-negative random variable can be expressed in terms of the cumulative distribution function F using,

$$2\int_0^\infty u(1 - F(u))\,du - \left(\int_0^\infty (1 - F(u))\,du\right)^2.$$

This expression can be used to calculate the variance in situations where the CDF, but not the density, can be conveniently expressed.

Characteristic Property

The second moment of a random variable attains the minimum value when taken around the first moment (i.e., mean) of the random variable, i.e. $\operatorname{argmin}_m \mathrm{E}\left(\left(X-m\right)^2\right)=\mathrm{E}(X)$. Conversely, if a continuous function φ satisfies $\operatorname{argmin}_m \mathrm{E}(\varphi(X-m))=\mathrm{E}(X)$ for all random variables *X*, then it is necessarily of the form $\varphi(x)=ax^2+b$, where $a>0$. This also holds in the multidimensional case.

Units of Measurement

Unlike expected absolute deviation, the variance of a variable has units that are the square of the units of the variable itself. For example, a variable measured in meters will have a variance measured in meters squared. For this reason, describing data sets via their standard deviation or root mean square deviation is often preferred over using the variance. In the dice example the standard deviation is $\sqrt{2.9\approx1.7}$, slightly larger than the expected absolute deviation of 1.5.

The standard deviation and the expected absolute deviation can both be used as an indicator of the "spread" of a distribution. The standard deviation is more amenable to algebraic manipulation than the expected absolute deviation, and, together with variance and its generalization covariance, is used frequently in theoretical statistics; however the expected absolute deviation tends to be more robust as it is less sensitive to outliers arising from measurement anomalies or an unduly heavy-tailed distribution.

Approximating the Variance of a Function

The delta method uses second-order Taylor expansions to approximate the variance of a function of one or more random variables. For example, the approximate variance of a function of one variable is given by,

$$Var[f(X)]\approx\left(f'(\mathrm{E}[X])\right)^2 Var[X]$$

provided that *f* is twice differentiable and that the mean and variance of *X* are finite.

Population Variance and Sample Variance

Real-world observations such as the measurements of yesterday's rain throughout the day typically cannot be complete sets of all possible observations that could be made. As such, the variance calculated from the finite set will in general not match the variance that would have been calculated from the full population of possible observations. This means that one estimates the mean and variance that would have been calculated from an omniscient set of observations by using an estimator equation. The estimator is a function of the sample of n observations drawn without observational bias from the whole population of potential observations. In this example that sample would be the set of actual measurements of yesterday's rainfall from available rain gauges within the geography of interest.

The simplest estimators for population mean and population variance are simply the mean and variance of the sample, the sample mean and (uncorrected) sample variance – these are consistent estimators (they converge to the correct value as the number of samples increases), but can be improved. Estimating the population variance by taking the sample's variance is close to optimal in general, but can be improved in two ways. Most simply, the sample variance is computed as an average of squared deviations about the (sample) mean, by dividing by n. However, using values other than n improves the estimator in various ways. Four common values for the denominator are n, $n - 1$, $n + 1$, and $n - 1.5$: n is the simplest (population variance of the sample), $n - 1$ eliminates bias, $n + 1$ minimizes mean squared error for the normal distribution, and $n - 1.5$ mostly eliminates bias in unbiased estimation of standard deviation for the normal distribution.

Firstly, if the omniscient mean is unknown (and is computed as the sample mean), then the sample variance is a biased estimator: it underestimates the variance by a factor of $(n - 1) / n$; correcting by this factor (dividing by $n - 1$ instead of n) is called Bessel's correction. The resulting estimator is unbiased, and is called the (corrected) sample variance or unbiased sample variance. For example, when $n = 1$ the variance of a single observation about the sample mean (itself) is obviously zero regardless of the population variance. If the mean is determined in some other way than from the same samples used to estimate the variance then this bias does not arise and the variance can safely be estimated as that of the samples about the (independently known) mean.

Secondly, the sample variance does not generally minimize mean squared error between sample variance and population variance. Correcting for bias often makes this worse: one can always choose a scale factor that performs better than the corrected sample variance, though the optimal scale factor depends on the excess kurtosis of the population, and introduces bias. This always consists of scaling down the unbiased estimator (dividing by a number larger than $n - 1$), and is a simple example of a shrinkage estimator: one "shrinks" the unbiased estimator towards zero. For the normal distribution, dividing by $n + 1$ (instead of $n - 1$ or n) minimizes mean squared error. The resulting estimator is biased, however, and is known as the biased sample variation.

Population Variance

In general, the population variance of a finite population of size N with values x_i is given by,

$$\sigma^2 = \frac{1}{N}\sum_{i=1}^{N}(x_i - \mu)^2 = \frac{1}{N}\sum_{i=1}^{N}\left(x_i^2 - 2\mu x_i + \mu^2\right)$$

$$= \left(\frac{1}{N} \sum_{i=1}^{N} x_i^2 \right) - 2\mu \left(\frac{1}{N} \sum_{i=1}^{N} x_i \right) + \mu^2$$

$$= \left(\frac{1}{N} \sum_{i=1}^{N} x_i^2 \right) - \mu^2$$

where the population mean is,

$$\mu = \frac{1}{N} \sum_{i=1}^{N} x_i.$$

The population variance can also be computed using,

$$\sigma^2 = \frac{1}{N^2} \sum_{i<j} \left(x_i - x_j \right)^2 = \frac{1}{2N^2} \sum_{i,j=1}^{N} \left(x_i - x_j \right)^2.$$

This is true because,

$$\begin{aligned} \frac{1}{2N^2} \sum_{i,j=1}^{N} \left(x_i - x_j \right)^2 &= \frac{1}{2N^2} \sum_{i,j=1}^{N} \left(x_i^2 - 2x_i x_j + x_j^2 \right) \\ &= \frac{1}{2N} \sum_{j=1}^{N} \left(\frac{1}{N} \sum_{i=1}^{N} x_i^2 \right) - \left(\frac{1}{N} \sum_{i=1}^{N} x_i \right) \left(\frac{1}{N} \sum_{j=1}^{N} x_j \right) \\ &\quad + \frac{1}{2N} \sum_{i=1}^{N} \left(\frac{1}{N} \sum_{j=1}^{N} x_j^2 \right) \\ &= \frac{1}{2} \left(\sigma^2 + \mu^2 \right) - \mu^2 + \frac{1}{2} \left(\sigma^2 + \mu^2 \right) \\ &= \sigma^2 \end{aligned}$$

The population variance matches the variance of the generating probability distribution. In this sense, the concept of population can be extended to continuous random variables with infinite populations.

Sample Variance

In many practical situations, the true variance of a population is not known *a priori* and must be computed somehow. When dealing with extremely large populations, it is not possible to count every object in the population, so the computation must be performed on a sample of the population. Sample variance can also be applied to the estimation of the variance of a continuous distribution from a sample of that distribution.

We take a sample with replacement of n values $Y_1, \ldots, Y_n$ from the population, where $n < N$, and estimate the variance on the basis of this sample. Directly taking the variance of the sample data gives the average of the squared deviations:

$$\sigma_Y^2 = \frac{1}{n}\sum_{i=1}^{n}\left(Y_i - \overline{Y}\right)^2 = \left(\frac{1}{n}\sum_{i=1}^{n}Y_i^2\right) - \overline{Y}^2 = \frac{1}{n^2}\sum_{i,j:i<j}\left(Y_i - Y_j\right)^2.$$

Here, $\overline{Y}$ denotes the sample mean:

$$\overline{Y} = \frac{1}{n}\sum_{i=1}^{n}Y_i.$$

Since the Y_i are selected randomly, both $\overline{Y}$ and σ_Y^2 are random variables. Their expected values can be evaluated by averaging over the ensemble of all possible samples $\{Y_i\}$ of size n from the population. For σ_Y^2 this gives:

$$\begin{aligned}
\mathrm{E}\left[\sigma_Y^2\right] &= \mathrm{E}\left[\frac{1}{n}\sum_{i=1}^{n}\left(Y_i - \frac{1}{n}\sum_{j=1}^{n}Y_j\right)^2\right] \\
&= \frac{1}{n}\sum_{i=1}^{n}\mathrm{E}\left[Y_i^2 - \frac{2}{n}Y_i\sum_{j=1}^{n}Y_j + \frac{1}{n^2}\sum_{k=1}^{n}Y_j\right] \\
&= \frac{1}{n}\sum_{i=1}^{n}\left[\frac{n-2}{n}\mathrm{E}\left[Y_i^2\right] - \frac{2}{n}\sum_{j\neq 1}^{n}\mathrm{E}\left[Y_iY_j\right] + \frac{1}{n^2}\sum_{j=1}^{n}\sum_{j=1}^{n}\mathrm{E}\left[Y_jY_k\right] + \frac{1}{n^2}\sum_{j=1}^{n}\mathrm{E}\left[Y_j^2\right]\right] \\
&= \frac{1}{n}\sum_{i=1}^{n}\left[\frac{n-2}{n}\left(\sigma^2+\mu^2\right) - \frac{2}{n}(n-1)\mu^2 + \frac{1}{n^2}n(n-1)\mu^2 + \frac{1}{n}\left(\sigma^2+\mu^2\right)\right] \\
&= \frac{n-1}{n}\sigma^2
\end{aligned}$$

Hence σ_Y^2 gives an estimate of the population variance that is biased by a factor of $\frac{n-1}{n}$. For this reason, σ_Y^2 is referred to as the *biased sample variance*. Correcting for this bias yields the *unbiased sample variance*:

$$s^2 = \frac{n}{n-1}\sigma_Y^2 = \frac{n}{n-1}\left(\frac{1}{n}\sum_{i=1}^{n}\left(Y_i - \overline{Y}\right)^2\right) = \frac{1}{n-1}\sum_{i=1}^{n}\left(Y_i - \overline{Y}\right)^2$$

Either estimator may be simply referred to as the *sample variance* when the version can be determined by context. The same proof is also applicable for samples taken from a continuous probability distribution.

The use of the term $n - 1$ is called Bessel's correction, and it is also used in sample covariance and the sample standard deviation (the square root of variance). The square root is a concave function and thus introduces negative bias (by Jensen's inequality), which depends on the distribution, and thus the corrected sample standard deviation (using Bessel's correction) is biased. The unbiased estimation of standard deviation is a technically involved problem, though for the normal distribution using the term $n - 1.5$ yields an almost unbiased estimator.

The unbiased sample variance is a U-statistic for the function $f(y_1, y_2) = (y_1 - y_2)^2/2$, meaning that it is obtained by averaging a 2-sample statistic over 2-element subsets of the population.

Distribution of the Sample Variance

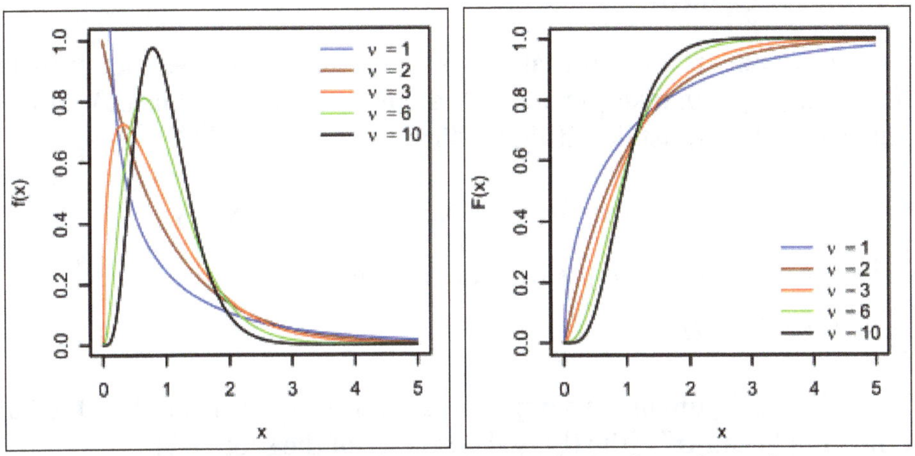

Distribution and cumulative distribution of S^2/σ^2, for various values of $v = n - 1$, when the y_i are independent normally distributed.

Being a function of random variables, the sample variance is itself a random variable, and it is natural to study its distribution. In the case that Y_i are independent observations from a normal distribution, Cochran's theorem shows that s^2 follows a scaled chi-squared distribution:

$$(n-1)\frac{s^2}{\sigma^2} \sim \chi^2_{n-1}.$$

As a direct consequence, it follows that,

$$\mathrm{E}(s^2) = \mathrm{E}\left(\frac{\sigma^2}{n-1}\chi^2_{n-1}\right) = \sigma^2,$$

and

$$Var[s^2] = Var\left(\frac{\sigma^2}{n-1}\chi^2_{n-1}\right) = \frac{\sigma^4}{(n-1)^2}Var\left(\chi^2_{n-1}\right) = \frac{2\sigma^4}{n-1}.$$

If the Y_i are independent and identically distributed, but not necessarily normally distributed, then,

$$\mathrm{E}[s^2] = \sigma^2, \quad Var[s^2] = \frac{\sigma^4}{n}\left((\kappa - 1) + \frac{2}{n-1}\right) = \frac{1}{n}\left(\mu_4 - \frac{n-3}{n-1}\sigma^4\right),$$

where κ is the kurtosis of the distribution and μ_4 is the fourth central moment.

If the conditions of the law of large numbers hold for the squared observations, s^2 is a consistent estimator of σ^2. One can see indeed that the variance of the estimator tends asymptotically to zero. An asymptotically equivalent formula was given in Kenney and Keeping, Rose and Smith, and Weisstein.

Samuelson's Inequality

Samuelson's inequality is a result that states bounds on the values that individual observations in a sample can take, given that the sample mean and (biased) variance have been calculated. Values must lie within the limits $\bar{y} \pm \sigma_Y (n-1)^{1/2}$.

Relations with the Harmonic and Arithmetic Means

It has been shown that for a sample $\{y_i\}$ of real numbers,

$$\sigma_y^2 \le 2y_{\max}(A - H),$$

where $y_{\max}$ is the maximum of the sample, A is the arithmetic mean, H is the harmonic mean of the sample and σ_y^2 is the (biased) variance of the sample.

This bound has been improved, and it is known that variance is bounded by,

$$\sigma_y^2 \le \frac{y_{\max}(A - H)(y_{\max} - A)}{y_{\max} - H},$$

$$\sigma_y^2 \ge \frac{y_{\min}(A - H)(A - y_{\min})}{H - y_{\min}},$$

where $y_{\min}$ is the minimum of the sample.

Tests of Equality of Variances

Testing for the equality of two or more variances is difficult. The F test and chi square tests are both adversely affected by non-normality and are not recommended for this purpose.

Several non parametric tests have been proposed: these include the Barton–David–Ansari–Freund–Siegel–Tukey test, the Capon test, Mood test, the Klotz test and the Sukhatme test. The Sukhatme test applies to two variances and requires that both medians be known and equal to zero. The Mood, Klotz, Capon and Barton–David–Ansari–Freund–Siegel–Tukey tests also apply to two variances. They allow the median to be unknown but do require that the two medians are equal.

The Lehmann test is a parametric test of two variances. Of this test there are several variants known. Other tests of the equality of variances include the Box test, the Box–Anderson test and the Moses test.

Resampling methods, which include the bootstrap and the jackknife, may be used to test the equality of variances.

Moment of Inertia

The variance of a probability distribution is analogous to the moment of inertia in classical mechanics of a corresponding mass distribution along a line, with respect to rotation about its center of mass. It is because of this analogy that such things as the variance are called *moments* of probability distributions. The covariance matrix is related to the moment of inertia tensor for multivariate distributions. The moment of inertia of a cloud of n points with a covariance matrix of Σ is given by,

$$I = n(\mathbf{1}_{3\times 3}\, tr(\Sigma) - \Sigma).$$

This difference between moment of inertia in physics and in statistics is clear for points that are gathered along a line. Suppose many points are close to the x axis and distributed along it. The covariance matrix might look like,

$$\Sigma = \begin{bmatrix} 10 & 0 & 0 \\ 0 & 0.1 & 0 \\ 0 & 0 & 0.1 \end{bmatrix}.$$

That is, there is the most variance in the x direction. Physicists would consider this to have a low moment *about* the x axis so the moment-of-inertia tensor is,

$$I = n\begin{bmatrix} 0.2 & 0 & 0 \\ 0 & 10.1 & 0 \\ 0 & 0 & 10.1 \end{bmatrix}.$$

Semivariance

The *semivariance* is calculated in the same manner as the variance but only those observations that fall below the mean are included in the calculation:

$$\text{Semivariance} = \frac{1}{n} \sum_{i:x_i<\mu} (x_i - \mu)^2$$

It is sometimes described as a measure of downside risk in an investments context. For skewed distributions, the semivariance can provide additional information that a variance does not.

For inequalities associated with the semivariance.

Generalizations

For Complex Variables

If x is a scalar complex-valued random variable, with values in $\mathbb{C}$, then its variance is $\mathrm{E}\left[(x-\mu)(x-\mu)^*\right]$, where x^* is the complex conjugate of x. This variance is a real scalar.

For Vector-valued Random Variables

As a Matrix

If X is a vector-valued random variable, with values in $\mathbb{R}^n$, and thought of as a column vector, then a natural generalization of variance is $\mathrm{E}\left[(X-\mu)(X-\mu)^{\mathrm{T}}\right]$, where $\mu=\mathrm{E}(X)$ and X^{T} is the transpose of X and so is a row vector. The result is a positive semi-definite square matrix, commonly referred to as the variance-covariance matrix (or simply as the *covariance matrix*).

If X is a vector- and complex-valued random variable, with values in $\mathbb{C}^n$, then the covariance matrix is $\mathrm{E}\left[(X-\mu)(X-\mu)^{\dagger}\right]$, where $X^{\dagger}$ is the conjugate transpose of X This matrix is also positive semi-definite and square.

As a Scalar

Another generalization of variance for vector-valued random variables X, which results in a scalar value rather than in a matrix, is the generalized variance $\det(C)$, the determinant of the covariance matrix. The generalized variance can be shown to be related to the multidimensional scatter of points around their mean.

A different generalization is obtained by considering the Euclidean distance between the random variable and its mean. This results in $\mathrm{E}\left[(X-\mu)^{\mathrm{T}}(X-\mu)\right]=\mathrm{tr}(C)$, which is the trace of the covariance matrix.

Coefficient of Variance

The coefficient of variation (CV) is a statistical measure of the dispersion of data points in a data series around the mean. The coefficient of variation represents the ratio of the standard deviation to the mean, and it is a useful statistic for comparing the degree of variation from one data series to another, even if the means are drastically different from one another.

The coefficient of variation shows the extent of variability of data in a sample in relation to the mean of the population. In finance, the coefficient of variation allows investors to determine how much volatility, or risk, is assumed in comparison to the amount of return expected from investments. Ideally, the coefficient of variation formula should result in a lower ratio of the standard deviation to mean return, meaning the better risk-return trade-off. Note that if the expected return in the denominator is negative or zero, the coefficient of variation could be misleading.

The coefficient of variation is helpful when using the risk/reward ratio to select investments. For example, an investor who is risk-averse may want to consider assets with a historically low degree of volatility and a high degree of return, in relation to the overall market or its industry. Conversely, risk-seeking investors may look to invest in assets with a historically high degree of volatility.

While most often used to analyze dispersion around the mean, quartile, quintile, or decile CVs can also be used to understand variation around the median or 10th percentile, for example.

Coefficient of Variation Formula

Below is the formula for how to calculate the coefficient of variation:

$$CV = \frac{\sigma}{\mu}$$

where,

σ =standard deviation

μ =mean

Please note that if the expected return in the denominator of the coefficient of variation formula is negative or zero, the result could be misleading.

Coefficient of Variation in Excel

The coefficient of variation formula can be performed in Excel by first using the standard deviation function for a data set. Next, calculate the mean using the Excel function provided. Since the coefficient of variation is the standard deviation divided by the mean, divide the cell containing the standard deviation by the cell containing the mean.

Example of Coefficient of Variation for Selecting Investments

For example, consider a risk-averse investor who wishes to invest in an exchange-traded fund (ETF), which is a basket of securities that tracks a broad market index. The investor selects the SPDR S&P 500 ETF, Invesco QQQ ETF, and the iShares Russell

2000 ETF. Then, he analyzes the ETFs' returns and volatility over the past 15 years and assumes the ETFs could have similar returns to their long-term averages.

For illustrative purposes, the following 15-year historical information is used for the investor's decision:

- SPDR S&P 500 ETF has an average annual return of 5.47% and a standard deviation of 14.68%. SPDR S&P 500 ETF's coefficient of variation is 2.68.
- Invesco QQQ ETF has an average annual return of 6.88% and a standard deviation of 21.31%. QQQ's coefficient of variation is 3.09.
- iShares Russell 2000 ETF has an average annual return of 7.16% and a standard deviation of 19.46%. IWM's coefficient of variation is 2.72.

Based on the approximate figures, the investor could invest in either the SPDR S&P 500 ETF or the iShares Russell 2000 ETF, since the risk/reward ratios are comparatively the same and indicate a better risk-return trade-off than the Invesco QQQ ETF.

POOLED VARIANCE

Pooled Variance/Change is the weighted normal for assessing the fluctuations of two autonomous variables where the mean can differ between tests however the genuine difference continues as before.

Problem Statement:

Compute the Pooled Variance of the numbers 1, 2, 3, 4 and 5.

Solution:

Step 1:

Decide the normal (mean) of the given arrangement of information by including every one of the numbers then gap it by the aggregate include of numbers given the information set.

$$Mean = \frac{1+2+3+4+5}{5} = \frac{15}{5} = 3$$

Step 2:

At that point, subtract the mean worth with the given numbers in the information set.

$$\Rightarrow (1-3),(2-3),(3-3),(4-3),(5-3) \Rightarrow -2,-1,0,1,2$$

Step 3:

Square every period's deviation to dodge the negative numbers.

$$\Rightarrow (-2)^2,(-1)^2,(0)^2,(1)^2,(2)^2 \Rightarrow 4,1,0,1,4$$

Step 4:

Now discover Standard Deviation utilizing the underneath equation.

$$S = \sqrt{\frac{\Sigma X - M^2}{n-1}}$$

$$\text{Standard Deviation} = \frac{\sqrt{10}}{\sqrt{4}} = 1.58113.$$

Step 5:

$$Pooled\,Variance\ (r) = \frac{((\text{aggregate check of numbers -1}) \times \text{Var})}{(\text{aggregate tally of numbers- 1})}$$

$$(r) = (5-1) \times \frac{2.5}{5-1},$$

$$= \frac{(4 \times 2.5)}{4} = 2.5$$

Hence, Pooled Variance (r) =2.5.

References

- Arithmetic-mean, statistics: tutorialspoint.com, Retrieved 16 July, 2019
- Quartile-deviation, business-mathematics-and-statistics-measures-of-central-tendency-and-dispersion: toppr.com, Retrieved 25 August, 2019
- Hippel, Paul T. von (2005). "Mean, Median, and Skew: Correcting a Textbook Rule". Journal of Statistics Education. 13 (2). doi:10.1080/10691898.2005.11910556
- Coefficientofvariation: investopedia.com, Retrieved 15 February, 2019
- Dodge, Yadolah (2003). The Oxford Dictionary of Statistical Terms. Oxford University Press. ISBN 978-0-19-920613-1
- Pooled-variance, statistics: tutorialspoint.com, Retrieved 09 May, 2019
- Frohne, I.; Hyndman, R.J. (2009). Sample Quantiles. R Project. ISBN 3-900051-07-0

4

Sampling Distributions

Sampling distribution refers to the probability distribution of data obtained from a large number of samples. Sampling distribution of mean, median, mode and standard deviation are studied within statistics. This chapter sheds light on the sampling distributions for an in-depth understanding of the subject.

Suppose we have a finite population and we draw all possible simple random samples of size n without replacement or with replacement. For each sample we calculate a statistic (sample mean $\overline{X}$ or proportion $\hat{p}$, etc.). All possible values of the statistic make a probability distribution which is called the sampling distribution. The number of all possible samples is usually very large and obviously the number of statistics (any function of the sample) will be equal to the number of samples if one and only one statistic is calculated from each sample. In fact, in practical situations, the sampling distribution has a very large number of values. The shape of the sampling distribution depends upon the size of the sample, the nature of the population and the statistic which is calculated from all possible simple random samples.

Standard Error

The standard deviation of a statistic is called the standard error of that statistic. If the statistic is $\overline{X}$, the standard deviation of all possible values of $\overline{X}$ is called the standard error of $\overline{X}$, which may be written as S.E.($\overline{X}$) or $\sigma_{\overline{X}}$. Similarly, if the sample statistic is proportion $\hat{p}$, the standard deviation of all possible values of $\hat{p}$ is called the standard error of $\hat{p}$ and is denoted by $\sigma_{\hat{p}}$ or S.E. ($\hat{p}$).

Sampling Distribution of $\overline{X}$

The probability distribution of all possible values of $\overline{X}$ calculated from all possible simple random samples is called the sampling distribution of $\overline{X}$. In brief, we shall call it the distribution of $\overline{X}$. The mean of this distribution is called the expected value of $\overline{X}$ and is written as $E\left(\overline{X}\right)$ $\sigma_{\overline{X}}$ The standard deviation (standard error) of this distribution is denoted by S.E.($\overline{X}$) or $\sigma_{\overline{X}}$ and the variance of $\overline{X}$ is denoted by Var($\overline{X}$) $Var\left(\overline{X}\right)$ or $\sigma^2_{\overline{X}}$. The distribution of $\overline{X}$ has some important properties:

- One important property of the distribution of $\overline{X}$ is that it is a normal distribution when the size of the sample is large. When the sample size n is more than 30, we call it a large sample size. The shape of the population distribution does not matter. The population may be normal or non-normal, the distribution of $\overline{X}$ is normal for n>30, but this is true when the number of samples is very large. As the distribution of random variable $\overline{X}$ is normal, $\overline{X}$ can be transformed into a standard normal variable Z where $Z = \dfrac{\overline{X} - \mu}{\sigma / \sqrt{n}}$. The distribution of $\overline{X}$ has a t-distribution when the population is normal and n>30. Diagram (a) shows the normal distribution and diagram (b) shows the t-distribution.

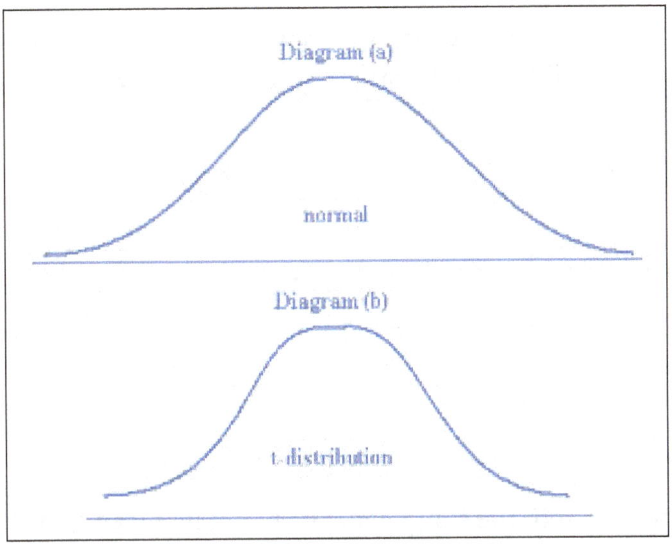

- The mean of the distribution of $\overline{X}$ is equal to the mean of the population. Thus $E(\overline{X}) = \mu_{\overline{X}} = \mu$ (population mean). This relation is true for small as well as large sample sizes in sampling without replacement and with replacement.
- The standard error (standard deviation) of $\overline{X}$ is related to the standard deviation of the population σ through the relations:

$$\text{S.E.}(\overline{X}) = \sigma_{\overline{X}} = \frac{\sigma}{\sqrt{n}}$$

This is true when population is infinite, which means NN is very large or the sampling is done with replacement from a finite or infinite population.

$$\text{S.E}(\overline{X}) = \sigma_{\overline{X}} = \frac{\sigma}{\sqrt{n}}\sqrt{\frac{N-n}{N-1}}$$

This is true when sampling is without replacement from a finite population. The above two equations between $\sigma_{\overline{X}}$ and σ are true both for small as well as large sample sizes.

Examples of Sampling Distribution

Draw all possible samples of size 2 without replacement from a population consisting of 3, 6, 9, 12, 15. Form the sampling distribution of sample means and verify the results.

- $E(\bar{X}) = \mu$
- $\text{Var}(\bar{X}) = \frac{\sigma^2}{n}\left(\frac{N-n}{N-1}\right)$

Solution:

We have population values 3, 6, 9, 12, 15, population size N=5 and sample size n=2. Thus, the number of possible samples which can be drawn without replacement is,

$$\binom{N}{n} = \binom{5}{2} = 10$$

Sample No.	Sample Values	Sample Mean $\bar{X}$	Sample No.	Sample Values	Sample Mean $\bar{X}$
1	3, 6	4.5	6	6, 12	9.0
2	3, 9	6.0	7	6, 15	10.5
3	3, 12	7.5	8	9, 12	10.5
4	3, 15	9.0	9	9, 15	12.0
5	6, 9	7.5	10	12, 15	13.5

The sampling distribution of the sample mean $\bar{X}$ and its mean and standard deviation are:

$\bar{X}$	f	$f(\bar{X})$	$\bar{X}f(\bar{X})$	$\bar{X}^2 f(\bar{X})$
4.5	1	1/10	4.5/10	20.25/10
6.0	1	1/10	6.0/10	36.00/10
7.5	2	2/10	15.0/10	112.50/10
9.0	2	2/10	18.0/10	162.00/10
10.5	2	2/10	21.0/10	220.50/10
12.0	1	1/10	12.0/10	144.00/10
13.5	1	1/10	13.5/10	182.25/10
Total	10	1	90/10	877.5/10

$$E(\bar{X}) = \Sigma \bar{X} f(\bar{X}) = \frac{90}{10} = 9$$

$$\text{Var}(\bar{X}) = \sum \bar{X}^2 f(\bar{X}) - \left[\sum \bar{X} f(\bar{X})\right]^2 = \frac{887.5}{10} - \left(\frac{90}{10}\right)^2 = 87.75 - 81 = 6.75$$

The mean and variance of the population are:

$\bar{X}$	3	6	9	12	15	$\Sigma X = 45$
X^2	9	36	81	144	225	$\Sigma = X^2 5$ $= 495$

$$\mu = \frac{\sum X}{N} = \frac{45}{5} = 9$$

and

$$\sigma^2 = \frac{\Sigma X^2}{N} - \left(\frac{\Sigma X}{N}\right)^2 = \frac{495}{5} - \left(\frac{45}{5}\right)^2 = 99 - 81 = 18$$

Verification:

- $E(\bar{X}) = \mu = 9$
- $Var(\bar{X}) = \frac{\sigma^2}{n}\left(\frac{N-n}{N-1}\right) = 182\left(\frac{5-2}{5-1}\right) = 6.75$

Example:

If random samples of size three are drawn without replacement from the population consisting of four numbers 4, 5, 5, 7. Find the sample mean $\bar{X}$ for each sample and make a sampling distribution of $\bar{X}$. Calculate the mean and standard deviation of this sampling distribution. Compare your calculations with the population parameters.

Solution:

We have population values 4, 5, 5, 7, population size N=4 and sample size n=3. Thus, the number of possible samples which can be drawn without replacement is,

$$\binom{N}{n} = \binom{4}{3} = 4$$

Sample No.	Sample Values	Sample Mean ($\bar{X}$)
1	4, 5, 5	14/3
2	4, 5, 7	16/3
3	4, 5, 7	16/3
4	5, 5, 7	17/3

The sampling distribution of the sample mean $\bar{X}$ and its mean and standard deviation are:

$\bar{X}$	f	$f(\bar{X})$	$\bar{X}f(\bar{X})$	$\bar{X}^2 f(\bar{X})$
14/3	1	1/4	14/12	196/36
16/3	2	2/4	32/12	512/36
17/3	1	1/4	17/12	289/36
Total	4	1	63/12	997/36

$$\mu_{\bar{X}} = \sum \bar{X} f(\bar{X}) = \frac{63}{12} = 5.25$$

$$\sigma_{\bar{X}} = \sqrt{\Sigma \bar{X}^2 f(\bar{X}) - \left[\Sigma \bar{X} f(\bar{X})\right]^2} = \sqrt{\frac{997}{36} - \left(\frac{63}{12}\right)^2} = 0.3632$$

The mean and standard deviation of the population are:

X	4	5	5	7	$\Sigma X = 21$
X^2	16	25	25	49	ΣX^2 $= 115$

$$\mu = \frac{\Sigma X}{N} = \frac{21}{4} = 5.25 \text{ and } \sigma^2 = \sqrt{\frac{\Sigma X^2}{N} - \left(\frac{\Sigma X}{N}\right)^2} = \sqrt{\frac{115}{4} - \left(\frac{21}{4}\right)^2} = 1.0897$$

$$\frac{\sigma}{\sqrt{n}}\sqrt{\frac{N-n}{N-1}} = \frac{1.0897}{\sqrt{3}}\sqrt{\frac{4-3}{4-1}} = 0.3632$$

Hence, $\mu_{\bar{X}} = \mu$ and $\sigma_{\bar{X}} = \dfrac{\sigma}{\sqrt{n}}\sqrt{\dfrac{N \quad n}{N-1}}$

Sampling Errors

Suppose we are interested in the value of a population parameter, the true value of which is θ but is unknown. The knowledge about θ can be obtained either from sample data or from population data. In both cases, there is a possibility of not reaching the true value of the parameter. The difference between the calculated value (from the sample data or from population data) and the true value of the parameter is called an error.

Thus, error is something which cannot be determined accurately if the population is large and the units of the population are to be measured. Suppose we are interested in finding the total production of wheat in Pakistan in a certain year. Sufficient funds and time are at our disposal and we want to get the 'true' figure of the production of wheat.

The maximum we can do is contact all the farmers, and suppose all the farmers cooperate completely and supply the information as honestly as possible. But the information supplied by the farmers will have errors in most cases, so we may not be able to identify the 'true' figure. In spite of all efforts, we shall be in the dark.

The calculated or observed figure may be good for all practical purposes but we can never claim that a true value of the parameter has been obtained. If the study of the units is based on counting, we can possibly get the true figure of the population parameter. There are two kinds of errors, (i) sampling errors or random errors and (ii) non-sampling errors.

Sampling Errors

Sampling errors occur due to the nature of sampling. The sample selected from the population is one of all possible samples. Any value calculated from the sample is based on the sample data and is called a sample statistic. The sample statistic may or may not be close to the population parameter. If the statistic is $\hat{\theta}$ and the true value of the population parameter is θ, then the difference $\hat{\theta} - \theta$ is called the sampling error. It is important to note that a statistic is a random variable and it may take any value.

A particular example of sampling error is the difference between the sample mean $\bar{X}$ and the population mean μ. Thus sampling error is also a random term. The population parameter is usually not known; therefore the sampling error is estimated from the sample data. The sampling error is due to the fact that a certain part of the population is incorporated in the sample. Obviously, one part of the population cannot give the true picture of the properties of the population. But one should not get the impression that a sample always gives a result which is full of errors. We can design a sample and collect sample data in a manner so that sampling errors are reduced. Sampling errors can be reduced by the following methods: (1) by increasing the size of the sample (2) by stratification.

Reducing Sampling Errors

- Increasing the size of the sample: The sampling error can be reduced by increasing the sample size. If the sample size n is equal to the population size N, then the sampling error is zero.

- Stratification: When the population contains homogeneous units, a simple random sample is likely to be representative of the population. But if the population contains dissimilar units, a simple random sample may fail to be representative of all kinds of units in the population. To improve the result of the sample, the sample design is modified. The population is divided into different groups containing similar units, and these groups are called strata. From each group (stratum), a sub-sample is selected in a random manner. Thus all groups are represented in the sample and the sampling error is reduced. This method is called stratified-random sampling. The size of the sub-sample from each stratum is frequently in proportion to the size of the stratum.

Suppose a population consists of 1000 students, out of which 600 are intelligent and 400 are unintelligent. We are assuming here that we do have much information about the population. A stratified sample of size n = 100 is to be selected. The size of the stratum is denoted by N_1 and N_2 respectively, and the size of the samples from each stratum may be denoted by n_1 and n_2. It is written as:

Stratum #	Size of stratum	Size of sample from each stratum
1	$N_1 = 600$	$n_1 = \frac{n \times N_1}{N} = \frac{100 \times 600}{1000} = 60$
2	$N_2 = 400$	$n_2 = \frac{n \times N_2}{N} = \frac{100 \times 400}{1000} = 40$
3	$N_1 + N_2 = N = 1000$	$n_1 + n_2 = n = 100$

The size of the sample from each stratum has been calculated according to the size of the stratum. This is called proportional allocation. In the above sample design, the sampling fraction in the population is $\frac{n}{N} = \frac{100}{1000} = \frac{1}{10}$ and the sampling fraction in both the strata is also $\frac{1}{10}$. Thus this design is also called a fixed sampling fraction. This modified sample & sign is frequently used in sample surveys. But this design requires some prior information about the units of the population, and the population is divided into different strata based on this information. If the prior information is not available then the stratification is not applicable.

The size of the sample from each stratum has been calculated according to the size of the stratum. This is called proportional allocation. In the above sample design, the sampling fraction in the population is $\frac{n}{N} = \frac{100}{1000} = \frac{1}{10}$ and the sampling fraction in both the strata is also $\frac{1}{10}$. Thus this design is also called a fixed sampling fraction. This modified sample & sign is frequently used in sample surveys. But this design requires some prior information about the units of the population, and the population is divided into different strata based on this information. If the prior information is not available then the stratification is not applicable.

Non Sampling Errors

There are certain sources of errors which occur both in sample surveys as well as in the complete enumeration. These errors are common in nature. Suppose we study each and every unit of the population. The population parameter under study is the population mean, and the 'true' value of the parameter is μ which is unknown. We hope to get the value of μ by a complete count of all the units of the population. We get a value called the 'calculated' or 'observed' value of the population mean. This observed value may be denoted by μ_{cal}. The difference between μ_{cal} and μ (true) is called a non-sampling error.

Even if we study the population units under ideal conditions, there may still be a difference between the observed value of the population mean and the true value of the population mean. Non-sampling errors may occur due to many reasons. Some of them are:

- The units of the population may not be defined properly. Suppose we have to carry out a study about the skilled labor force in our country. Who is a skilled person? Some people do more than one job. Some do the administrative jobs as well as the technical jobs. Some are skilled but they are working in an unskilled position. Thus it is important to clearly define the units of the population, otherwise there will be non-sampling errors both in the population count and the sample study.
- There may be a poor response on the part of respondents, and they do not supply correct information about their income, their children, their age and property, etc. These errors are likely to be of a high magnitude in population study than the sample study. To reduce these errors the respondents are to be persuaded.
- Data collection is subject to human error. The enumerators may be careless or they may not be able to maintain uniformity from place to place. The data may not be collected properly from the population or from the sample. These errors are likely to be more serious in the population data than the sample data.
- Another serious error is due to bias. Bias means an error on the part of the enumerator or the respondent when the data is being collected, and it may be intentional or unintentional. An enumerator may not be capable of reporting the correct data. If they have to report about the condition of crops in different areas after heavy rainfalls, their assessments may be biased due to lack of training or they may be inclined to give inaccurate reports. Bias is a serious error and cannot be reduced by increasing the sample size. Bias may be present in the sample study as well as the population study.

Importance of Sampling Distribution in Research Methodology

Some important sampling distributions, which are commonly used, are:

- Sampling distribution of mean.
- Sampling distribution of proportion.
- Student's 't' distribution.

- F distribution.
- Chi-square distribution.

A brief mention of each one of these sampling distribution will be helpful:

- Sampling distribution of mean: Sampling distribution of mean refers to the probability distribution of all the possible means of random samples of a given size that we take from a population. If samples are taken from a normal population, N dm,s p i, the sampling distribution of mean would also be normal with mean mx = m and standard deviation = s p n, where m is the mean of the population, s p is the standard deviation of the population and n means the number of items in a sample. But when sampling is from a population which is not normal (may be positively or negatively skewed), even then, as per the central limit theorem, the sampling distribution of mean tends quite closer to the normal distribution, provided the number of sample items is large i.e., more than 30. In case we want to reduce the sampling distribution of mean to unit normal distribution i.e., N (0,1), we can write the normal variate Formula for the sampling distribution of mean. This characteristic of the sampling distribution of mean is very useful in several decision situations for accepting or rejection of hypotheses.
- Sampling distribution of proportion: Like sampling distribution of mean, we can as well have a sampling distribution of proportion. This happens in case of statistics of attributes. Assume that we have worked out the proportion of defective parts in large number of samples, each with say 100 items, that have been taken from an infinite population and plot a probability distribution of the said proportions, we obtain what is known as the sampling distribution of the said proportions, we obtain what is known as the sampling distribution of proportion. Usually the statistics of attributes correspond to the conditions of a binomial distribution that tends to become normal distribution as n becomes larger and larger. If p represents the proportion of defectives i.e., of successes and q the proportion of non- defectives i.e., of failures (or q = 1 – p) and if p is treated as a random variable, then the sampling distribution of proportion of successes has a mean = p with standard deviation = Formula where n is the sample size. Presuming the binomial distribution approximating the normal distribution for large n, the normal variate of the sampling distribution of proportion z = Formula where $p (pronounced as p-hat) is the sample proportion of successes, can be used for testing of hypotheses.
- Student's t-distribution: When population standard deviation Formula is not known and the sample is of a small size b i.e., n < 30 g , we use t distribution for the sampling distribution of mean and workout t variable as:

  $$t = \left(\bar{X} - \mu\right) / \left(\sigma_s / \sqrt{n}\right)$$

where $\sigma_s = \sqrt{\frac{\Sigma\left(X_i - \bar{X}\right)^2}{n} - 1}$

i.e., the sample standard deviation. t-distribution is also symmetrical and is very close to the distribution of standard normal variate, z, except for small values of n. The variable t differs from z in the sense that we use sample standard deviation s s b g in the calculation of t, whereas we use standard deviation of population s p d i in the calculation of z. There is a different t distribution for every possible sample size i.e., for different degrees of freedom. The degrees of freedom for a sample of size n is n – 1. As the sample size gets larger, the shape of the t distribution becomes apporximately equal to the normal distribution. In fact for sample sizes of more than 30, the t distribution is so close to the normal distribution that we can use the normal to approximate the t-distribution. But when n is small, the t-distribution is far from normal but when n a, t-distribution is identical with normal distribution. The t-distribution tables are available which give the critical values of t for different degrees of freedom at various levels of significance. The table value of t for given degrees of freedom at a certain level of significance is compared with the calculated value of t from the sample data, and if the latter is either equal to or exceeds, we infer that the null hypothesis cannot be accepted.

- F distribution: F ratio is computed in a way that the larger variance is always in the numerator. Tables have been prepared for F distribution that give critical values of F for various values of degrees of freedom for larger as well as smaller variances. The calculated value of F from the sample data is compared with the corresponding table value of F and if the former is equal to or exceeds the latter, then we infer that the null hypothesis of the variances being equal cannot be accepted. We shall make use of the F ratio in the context of hypothesis testing and also in the context of ANOVA technique.

- Chi-square Formula distribution: Chi-square distribution is encountered when we deal with collections of values that involve adding up squares. Variances of samples require us to add a collection of squared quantities and thus have distributions that are related to chi-square distribution. If we take each one of a collection of sample variances, divide them by the known population variance and multiply these quotients by (n – 1), where n means the number of items in the sample, we shall obtain a chi-square distribution. Thus, Formula would have the same distribution as chi-square distribution with (n – 1) degrees of freedom. Chi-square distribution is not symmetrical and all the values are positive. One must know the degrees of freedom for using chi-square distribution. This distribution may also be used for judging the significance of difference between observed and expected frequencies and also as a test of goodness of fit. The generalised shape of c 2 distribution depends upon the d.f. and the c 2 value is worked out as under:

$$\chi^2 = \sum_{i=1}^{k} \frac{(O_i - E_i)^2}{E_i}.$$

SAMPLING DISTRIBUTIONS RELATED TO THE NORMAL DISTRIBUTION

Theorem: Let $X_1, X_2, \ldots, X_n$ L be a random sample of size n from a normal distribution with mean μ and variance σ^2. Then,

$$\bar{X} = \frac{1}{n}\sum_{i=1}^{n} X_i$$

is normally distributed with mean $\mu_{\bar{X}} = \mu$ and $\sigma^2_{\bar{X}} = \sigma^2/n, \bar{X} \sim N\left(\mu, \sigma^2/n\right)$.

Proof: Since the moment-generating function of each X is,

$$M_X(t) = \exp\left(\mu t + \sigma^2 t^2 / 2\right)$$

the moment-generating function of $\bar{X} = \frac{1}{n}\sum_{i=1}^{n} X_i$ is equal to,

$$M_{\bar{X}}(t) = E\left(e^{(t/n)\Sigma X_i}\right) = \prod_{i=1}^{n} M_{X_i}\left(\frac{t}{n}\right) = \left\{\exp\left[\mu\left(\frac{t}{n}\right) + \frac{\sigma^2 (t/n)^2}{2}\right]\right\}^n$$

$$= \exp\left[\mu t + \frac{\left(\sigma^2/n\right)t^2}{2}\right]$$

However, the moment-generating function uniquely determines the distribution of the random variable. Since this one is that associated with the normal distribution N(μ, σ^2/n), the sample mean $\bar{X}$ is N(μ, σ^2/n).

Theorem: Let $Z_1, Z_2, \ldots, Z_n$ have standard normal distributions, N(0, 1). If these random variables are mutually independent, then $W = Z_1^2, Z_2^2, \ldots, Z_n^2$ has a χ^2 distribution with n degrees of freedom.

Proof: The moment-generating function of W is given by,

$$M_w(t) = E(e^{tW}) = E(\exp\{t(Z_1^2 + Z_2^2 + ... + Z_n^2)\}) = M_{Z_1^2}(t) \times M_{Z_2^2}(t) \times ... \times M_{Z_n^2}(t)$$

Since Z_i^2 is a χ^2 distributed random variable with 1 degree of freedom, we have,

$$M_{Z_i^2}(t) = (1-2t)^{-1/2}, \quad i = 1,2,...,n.$$

Hence,

$$M_w(t) = (1-2t)^{-1/2} \times (1-2t)^{-1/2} \times ... \times (1-2t)^{-1/2} = (1-2t)^{-n/2}$$

The uniqueness of the moment-generating function implies that W is χ^2 (n).

Corollary: If $X_1, X_2, ..., X_n$ have mutually independent normal distributions $N(\mu_i, \sigma_i^2), i = 1,2,...,n,$, respectively, then the distribution of,

$$W = \sum_{i=1}^{n} \frac{(X_i - \mu_i)^2}{\sigma_i^2}$$

has a χ^2 distribution with n degrees of freedom.

Proof: We simply note that $Z_i = (X_i - \mu_i)/\sigma_i$ is N(0,1), i = 1, 2,...,n.

Theorem: Let $X_1, X_2, ..., X_n$ be a random sample of size n from a normal distribution, $N(\mu, \sigma^2)$,

$$\bar{X} = \frac{1}{n}\sum_{i=1}^{n} X_i$$

$$S^2 = \frac{1}{n-1}\sum_{i=1}^{n}(X_i - \bar{X})^2$$

Then,

- $\bar{X}$ and S^2 are independent.
- $\frac{(n-1)S^2}{\sigma^2} = \frac{\sum_{i=1}^{n}(X_i - \bar{X})^2}{\sigma^2}$ has a χ^2 distribution with (n – 1) degrees of freedom.

Proof:

$$\frac{\sum_{i=1}^{n}(X_i - \mu)^2}{\sigma^2} = \sum_{i=1}^{n}\left[\frac{(X_i - \bar{X}) + (\bar{X} - \mu)}{\sigma}\right]^2 = \sum_{i=1}^{n}\left(\frac{X_i - \bar{X}}{\sigma}\right)^2 + \frac{n(\bar{X} - \mu)^2}{\sigma^2}$$

since the cross-product term is equal to,

$$2\sum_{i=1}^{n}\frac{(\bar{X}-\mu)(X_i-\bar{X})}{\sigma^2}=\frac{2(\bar{X}-\mu)}{\sigma^2}\sum_{i=1}^{n}(X_i-\bar{X})=0$$

But $Y_i = (X_i - \mu)/\sigma$, $i = 1,2,\ldots,n$, are mutually independent standard normal variables. Hence $W = Y_1^2 + Y_2^2 + \ldots + Y_n^2$ is χ^2(n) χ^2 by Theorem. Moreover, since,

$\bar{X} \sim N(\mu,\sigma^2/n)$, then $Z^2 = \left(\dfrac{\bar{X}-\mu}{\sigma/\sqrt{n}}\right)^2 = \dfrac{n(\bar{X}-\mu)^2}{\sigma^2}$ is χ^2 thus,

$$W = \frac{(n-1)S^2}{\sigma^2} + Z^2.$$

However, $\overline{X}$ and S^2 are independent; thus Z^2 and S^2 are also independent. In the moment-generating function of W, this independence permits us to write,

$$E\left(e^{t[(n-1)S^2/\sigma^2+Z^2]}\right)=E\left(e^{t(n-1)S^2/\sigma^2}e^{tZ^2}\right)=E\left(e^{t(n-1)S^2/\sigma^2}\right)E\left(e^{tZ^2}\right)$$

Since W and Z^2 have χ^2 distributions, we can substitute their moment-generating functions to obtain,

$$(1-2t)^{-n/2}=E\left(e^{t(n-1)S^2/\sigma^2}\right)(1-2t)^{-1/2}$$

Equivalently, we have,

$$E\left(e^{t(n-1)S^2/\sigma^2}\right)=(1-2t)^{-(n-1)/2}, t<1/2$$

This, of course, is the moment-generating function of a χ^2 (n – 1) variable, and accordingly (n –)S^2/σ^2 has this distribution.

Theorem: If Z is a standard normal distribution, N(0,1), if U is a χ^2 distribution with v degrees of freedom, and if Z and U are independent, then:

$$T=\frac{Z}{\sqrt{U/v}}$$

has a t distribution with v degrees of freedom. Its p.d.f. is,

$$f(t)=\frac{\Gamma[(v+1)/2]}{\Gamma(v/2)}\frac{1}{\sqrt{\pi v}}\left(1+\frac{t^2}{v}\right)^{-(v+1)/2}, \quad -\infty<t<\infty$$

This distribution was discovered by W. S. Gosset when he was working for an Irish brewery. Because Gosset published under the pseudonym Student, this distribution is sometimes known as Student's t distribution.

Proof: Since Z and U are independent, the joint p.d.f. of Z and U is,

$$g(z,u)=\frac{1}{\sqrt{2\pi}}e^{-z^2/2}\frac{1}{\Gamma(v/2)2^{v/2}}u^{v/2-1}\,e^{-u/2}, \quad -\infty<z<\infty, 0<u<\infty$$

The distribution function F(t) = Pr(T ≤ t) of T is given by,

$$F(t)=\Pr\left(Z/\sqrt{U/v}\le t\right)=\Pr\left(Z\le t\sqrt{U/v}\right)=\int_0^{\infty}\int_{-\infty}^{t\sqrt{u/v}}g(z,u)dzdu$$

The p.d.f. of T is the derivative of the distribution function; so applying the Fundamental Theorem of Calculus to the inner integral we see that,

$$f(t)=F'(t)=\frac{1}{\sqrt{\pi}\Gamma(v/2)}\int_0^{\infty}\frac{e^{-(u/2)\left(t^2/v\right)}}{2^{(v+1)/2}}\sqrt{\frac{u}{v}}\,u^{v/2-1}e^{-u/2}du$$

$$=\frac{1}{\sqrt{\pi}\Gamma(v/2)}\int_0^{\infty}\frac{u^{(v+1)/2-1}}{2^{(v+1)/2}}e^{-(u/2)\left(1+t^2/v\right)}du$$

In the integral make the change of variables $y=\left(1+t^2/v\right)u$ so that $du/dy=1/\left(1+t^2/v\right)$. Thus we find that,

$$f(t)=\frac{\Gamma((v+1)/2)}{\Gamma(v/2)}\frac{1}{\sqrt{\pi v}}\left(1+\frac{t^2}{v}\right)^{-(v+1)/2}\int_0^{\infty}\frac{y^{(v+1)/2-1}}{\Gamma((v+1)/2)2^{(v+1)/2}}e^{-y/2}dy.$$

The integral is equal to 1 since the integrand is the p.d.f of a chi-square distribution with (v + 1) degrees of freedom. Thus the p.d.f. is as given in the theorem.

Note that the distribution of T is completely determined by the number v. Its p.d.f. is symmetrical with respect the vertical axis $t=0$ and is very similar to the graph of the p.d.f. of the standard normal distribution N(0,1). It can be shown that $E(T)=0$ for $v>1$ and $Var(T)=v/(v-2)$ for $v>2$[1] When $v=1$, the t distribution is the same as the standard Cauchy distribution in which the mean and the variance do not exist.

Theorem : If $X_1, X_2, \ldots, X_n$ denote a random sample from $N\left(\mu,\sigma^2\right)$, then:

$$\frac{\bar{X}-\mu}{S/\sqrt{n}}\sim t(n-1).$$

Proof: This follows from Theorem , since $\left(\bar{X}-\mu\right)/\left(\sigma/\sqrt{n}\right)\sim N(0,1)$ and by Theorem , $U=(n-1)S^2/\sigma^2\sim\chi^2(n-1)$ and $\bar{X}$ and S^2 are independent.

Example: Let $X_1,X_2,...,X_n$ and $Y_1,Y_2,...,Y_m$ be independent random samples from populations with respectively distributions $X^i \sim N(\mu_x,\sigma_x^2)$ and $Y_j \sim N(\mu_y,\sigma_y^2)$ The distributions of $\bar{X}$ and $\bar{Y}$ are $N(\mu_x,\sigma_x^2/n)$ and $N(\mu_y,\sigma_y^2/m)$ respectively. Since $\bar{X}$ and $\bar{Y}$ are independent, the distribution $\bar{X}-\bar{Y}$ is $N(\mu_x-\mu_y\sigma_x^2/n+\sigma_y^2/m)$ and

$$Z=\frac{(\bar{X}-\bar{Y})-(\mu_x-\mu_y)}{\sqrt{\sigma_x^2/n+\sigma_y^2/m}}\sim N(0,1)$$

Since $(n-1)S_x^2/\sigma_x^2 \sim \chi^2(n-1)$ and $(m-1)S_y^2/\sigma_y^2 \sim \chi^2(m-1)$, and both are independent,

$$U=\frac{(n-1)S_x^2}{\sigma_x^2}+\frac{(m-1)S_y^2}{\sigma_y^2}\sim \chi^2(n+m-2)$$

A random variable T with the t distribution having v = n + m – 2 degrees of freedom is given by,

$$T\frac{Z}{\sqrt{U/(n+m-2)}}=\frac{\left[(\bar{X}-\bar{Y})-(\mu_x-\mu_y)\right]/\sqrt{\sigma_x^2/n+\sigma_y^2/m}}{\sqrt{((n-1)S_x^2/\sigma_x^2+(m-1)S_y^2/\sigma_y^2)/(n+m-2)}}$$

In the statistical applications we sometimes assume that the two variances are the same, say $\sigma_x^2+\sigma_y^2=\sigma^2$ in which case,

$$T=\frac{(\bar{X}-\bar{Y})-(\mu_x-\mu_y)}{\sqrt{\{[(n-1)S_x^2+(m-1)S_y^2]/(n+m-2)\}[(1/n)+(1/m)]}}$$

and neither T nor its distribution dependent on σ^2.

Theorem: If U_1 and U_2 are independent chi-square variables with v_1 and v_2 degrees of freedom, respectively, then:

$$F=\frac{U_1/v_1}{U_2/V_2}$$

Has an F distribution with v1 and v2 degrees of freedom. Its p.d.f. is,

$$f(y)=\frac{\Gamma[(v_1+v_2)/2]}{\Gamma(v_1/2)\Gamma(v_2/2)}\left(\frac{v_1}{v_2}\right)^{V_1/2}y^{(v_1/2)-1}\left(1+\frac{v_1}{v_2}y\right)^{-(v_1+v_2)/2},0<y<\infty$$

Example: Let $X_1,X_2,...,X_n$ and $Y_1,Y_2,...,Y_m$ be independent random samples from populations with respectively distributions $X^i \sim N(\mu_x, \sigma_x^2)$ and $Y_j \sim N(\mu_y, \sigma_y^2)$. Since $(n-1)S_x^2/\sigma_x^2 \sim \chi^2(n-1)$ and $(m-1)S_y^2/\sigma_y^2 \sim \chi^2(m-1)$,

$$\frac{((n-1)s_x^2/\sigma_x^2)/(n-1)}{((m-1)s_y^2/\sigma_y^2)/(m-1)} = \frac{S_x^2\sigma_y^2}{S_y^2\sigma_x^2} \sim F((n-1),(m-1)).$$

SAMPLING DISTRIBUTION OF SAMPLE MEAN

Consider a population of N variates with mean μ and standard deviation σ, and draw all possible samples of r variates. Assume that the samples have been replaced before each drawing, so that the total number of different samples which can be drawn is the combination of N things taken r at a time, that is M = $\binom{N}{r}$. The mean of all these sample means $(\bar{Y}_1, \bar{Y}_2, \ldots\ldots, \bar{Y}_M)$ is denoted by $\mu_{\bar{Y}}$ and their standard deviation by $\sigma_{\bar{Y}}$, also known as the standard error of a mean. The mean of the sample means is the same as the mean of the parent population, μ , e.g,

$$\mu_{\bar{Y}} = \Sigma\bar{Y}_i / \mathrm{M} = \mu = \Sigma Y_i / N$$

The variance of the sample means ($\sigma\frac{2}{y}$) equals the variance of the parent (σ^2) population divided by the sample size (r) and multiplied by a factor f.

$$\sigma_{\bar{y}}^2 = \Sigma(\bar{Y}_i - \mu_{\bar{Y}})^2 / M = (\sigma^2 / r)\bullet \mathrm{f}$$

where, f = (N - r)/(N - 1)

Note that the standard error of a mean approaches the standard deviation of the parent population divided by the square root of the sample size, $\sigma_{\bar{Y}} = \sigma / \sqrt{r}$ for a large population (i.e., f approaches unity). The larger the size of a sample, the smaller the variance of the sample mean.

Consider samples taken from a normal population. Figure illustrates the relationship of the parent population (r = 1) with the sampling distributions of the means of samples of size r = 8 and r = 16.

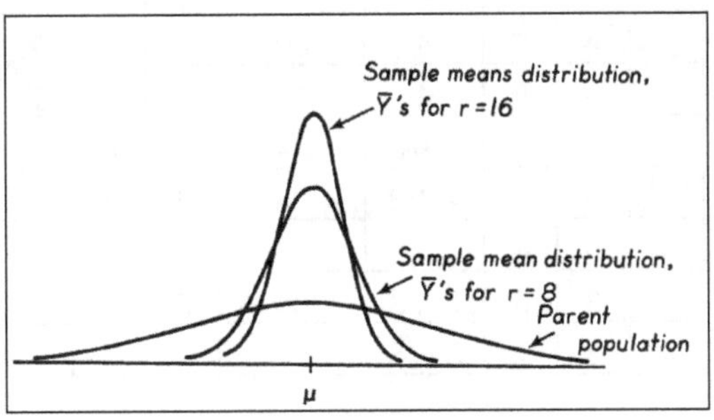

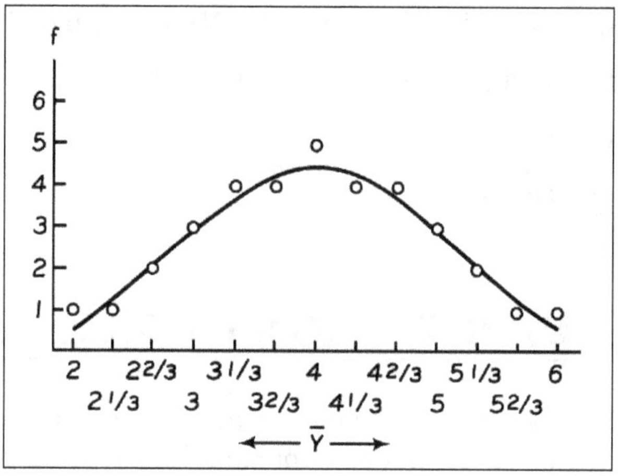

The relation of the frequencies of means for r = 3 from the population 1,2,3,4,5,6,7 and the normal distribution.

Even when the variates of the parent population are not normally distributed, the means generated by samples tend to be normally distributed. This can be illustrated by considering samples of size 3 from a simple non-normal population with variates 1,2,3,4,5,6, and 7. Table presents all possible sample means, and figure shows the frequency distribution of the means which approaches the normal frequency curve.

All possible different samples of size 3 from the population 1, 2, 3, 4, 5, 6, 7 with $\mu = 4$ and $\sigma = 2$.

Sample No.	Sample			$\bar{Y}$	Sample No.	Sample			
1	1	2	3	2	19	2	3	7	4
2	1	2	4	2 1/3	20	2	4	5	3 2/3
3	1	2	5	2 2/3	21	2	4	6	4
4	1	2	6	3	22	2	4	7	4 1/3
5	1	2	7	3 1/3	23	2	5	6	4 1/3
6	1	3	4	2 2/3	24	2	5	7	4 2/3
7	1	3	5	3	25	2	6	7	5
8	1	3	6	3 1/3	26	3	4	5	4
9	1	3	7	3 2/3	27	3	4	6	4 1/3
10	1	4	5	3 1/3	28	3	4	7	4 2/3
11	1	4	6	3 2/3	29	3	5	6	4 2/3
12	1	4	7	4	30	3	5	7	5
13	1	5	6	4	31	3	6	7	5 1/3
14	1	5	7	4 1/3	32	4	5	6	5
15	1	6	7	4 2/3	33	4	5	7	5 1/3
16	2	3	4	3	34	4	6	7	5 2/3
17	2	3	5	3 1/3	35	5	6	7	6
18	2	3	6	3 2/3					

The mean and standard deviation of the distribution of the sample means are:

$$\mu_{\bar{y}} = \frac{1}{35}(2+21/3+22/3+...+52/3+6) = 4 = \mu$$

$$\sigma_{\bar{y}}^2 = \frac{1}{35}\{(2-4)^2+(21/3-4)^2+...+(52/3-4)^2+(6-4)^2\}$$

$$= \frac{\sigma^2}{r} \bullet (\frac{N-r}{N-1}) = \frac{4}{3} \bullet (\frac{4}{6}) = \frac{8}{9}$$

$$\sigma_{\bar{y}} = \sqrt{8/9}$$

Note that in this particular case, we have used a simple population with only seven elements. Sample means from samples with increasing size, from a large population will more closely approach the normal curve. This tendency of sample means to approach a normal distribution with increasing sample size is called the central limit theorem.

Sampling Distribution of the Difference between Two Means

Above, we mentioned that the means of all possible samples of a given size (r1) drawn from a large population of Y's are approximately normally distributed with $\mu_{\bar{y}} = \mu_y$ and $\sigma_{\bar{y}}^2 = \sigma_y^2 / r_1$. Now consider drawing samples of size r2 from another large population, X's. The parameters of these sample means are also approximately normally distributed with $\mu_{\bar{x}} = \mu_x$ and $\sigma_{\bar{x}}^2 = \sigma_x^2 / r_2$. An additional approximately normal population is generated by taking differences between all possible means, $\bar{Y} - \bar{X} = \bar{d}$, with the parameters $\mu_{\bar{d}}$ and $\sigma_{\bar{d}}^2$,

$$\mu_{\bar{d}} = \mu_{\bar{y}} - \mu_{\bar{x}} = \mu_y - \mu_x$$

and

$$\sigma_{\bar{d}}^2 = \sigma_{\bar{y}}^2 + \sigma_{\bar{x}}^2 = \sigma_y^2 / r_1 + \sigma_x^2 / r_2$$

When the variances of the parent populations are equal,

$\sigma_y^2 = \sigma_x^2 (= \sigma^2)$ and sample sizes are the same, $r = r_1 = r_2$ then $\sigma_{\bar{d}}^2 = 2\sigma^2 / r$.

The square root of the variance of mean differences, $\sigma_{\bar{d}}$, is usually called the standard error of the difference between sample means. Figure diagrams the generation of a population of mean differences by repeated sampling from two populations of individual variates and indicates relationships among the parameters.

The relationships among the population parameters developed are important in statistical evaluation. With information about the parent population one can estimate parameters associated with a sample mean or the difference between two sample means.

Population Mean
$$\mu = \frac{\sum_{i=1}^{N} x_i}{N}$$ N = number of items in the population

Sample Mean
$$\overline{X} = \frac{\sum_{i=1}^{n} x_i}{n}$$ n = number of items in the sample

Relationships between parameters of a population of sample mean differences and parent populations.

Normal Approximation to Binomial

If the number of trials (n) is large the calculations become tedious. Since many practical problems involve large samples of repeated trials, it is important to have a more rapid method of finding binomial probabilities.

When $n > 30$, the sample is usually considered large. In this topic we will show how the normal distribution is used to approximate a binomial distribution for ease in the calculation of probabilities.

Since the normal frequency curve is always symmetric, whereas the binomial histogram is symmetric only when $p = q = 1/2$, it is clear that the normal curve is a better approximation of the binomial histogram if both p and q are equal to or nearly equal to 1/2. The more p and q differ from 1/2, the greater the number of trials are required for a close approximation. Figure shows how closely a normal curve can approximate a binomial distribution with $n = 10$ and $p = q = 1/2$. Figure illustrates a case where the normal distribution closely approximates the binomial when p is small but the sample size is large.

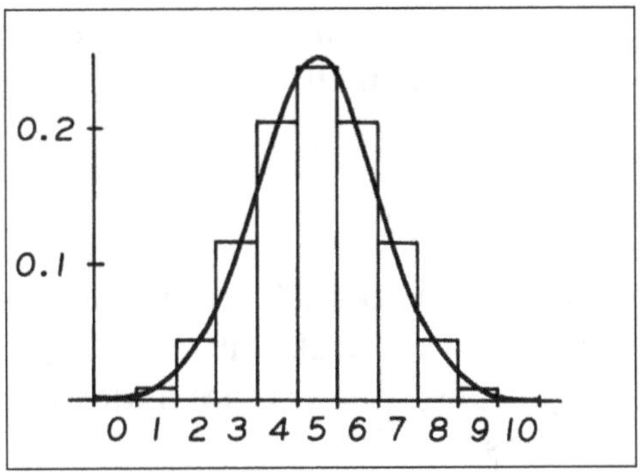

Binomial distribution for p = 0.5 and n = 10.

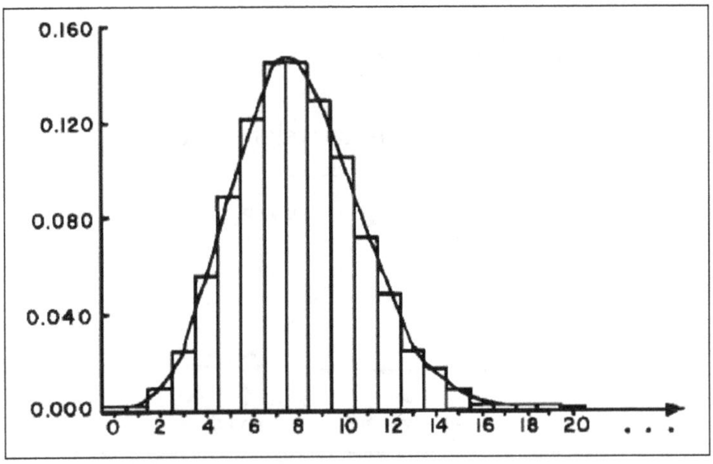

Binomial distribution for p = 0.08 and n = 100.

To use the normal curve to approximate discrete binomial probabilities, the area under the curve must include the area of the block of the histogram at any value of r, the number of occurrences under consideration. To include the block centered at r, the value of Y to be used in the normal curve equation for the normal deviate must be adjusted by adding 1/2 to, or subtracting 1/2 from the value of r. The calculation can be described by the following steps:

Step 1. Compute the mean and the standard deviation,

$$\mu = \text{np}, \quad \sigma = \sqrt{\text{npq}}$$

Step 2. In order to find the corresponding normal deviate (Y) for a given r, 1/2 must be either added to or subtracted from r to include the block centered at r.

$$\text{Y} = r - 1/2 \text{ or } \text{Y} = r + 1/2$$

Step 3. Standardize the normal deviate Y, by computing Z.

$$Z = (Y - np) / \sqrt{npq}$$

Step 4. The probability of the occurrence of a random standard normal deviate that is equal to or greater than, or equal to or smaller than Z.

Step 5. Compute the required probability. This depends on the nature of the problem and is illustrated by the four cases below.

Example. If 8% of a particular canned product is known to be underweight, what is the probability that a random sample of 100 cans will contain (a) 14 or more underweight cans (b) 4 or fewer underweight cans, (c) 5 or more underweight cans, (d) more than 4 but less than 15 underweight cans?

Step 1. μ = np = 100 (0.08) = 8.0,

$$\sigma = \sqrt{npq} = \sqrt{10(0.08)(0.92)} = 2.71$$

(a) To find the probability of 14 or more underweight cans.

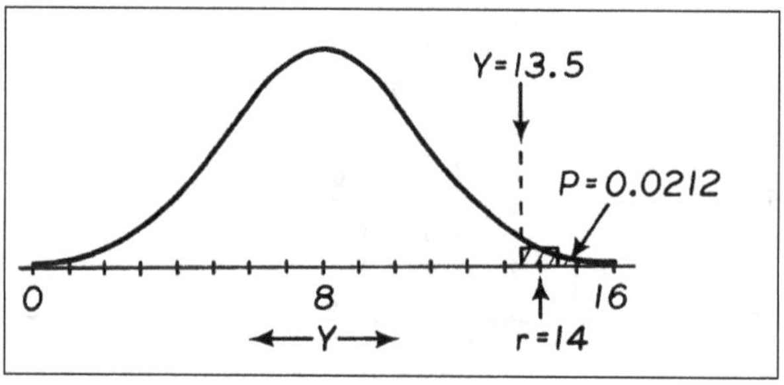

Step 2. Y = 5 - 1/2 = 14 - 1/2 = 13.5.

Step 3. Z = (13.5 - 8.0) / 2.71 = 2.03.

Step 4. $P(Z \geq 2.03) = 0.0212$.

Step 5. The required probability in this case is the one obtained from Step 4, 0.0212.

(b) To find the probability of 4 or fewer underweight cans,

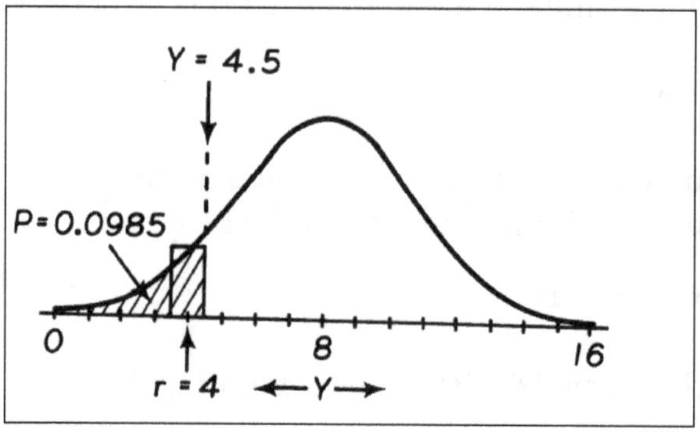

Step 2. Y = r + 1/2 = 4 + 0.5 = 4.5.

Step 3. Z = (4.5 - 8.0) / 2.71 = 1.29.

Step 4. Positive values for Z, i.e., for $Z \geq 0$. since the distribution is symmetrical about Z = 0, probabilities for negative values of Z are determined by ignoring the sign. Therefore, $P\ (Z \leq -1.29)\ =\ P\ (Z \leq 1.29) = 0.0985$.

Step 5. The required probability in this case is the one obtained from Step 4, 0.0985, or about 10%.

(c) To find the probability of 5 or more underweight cans,

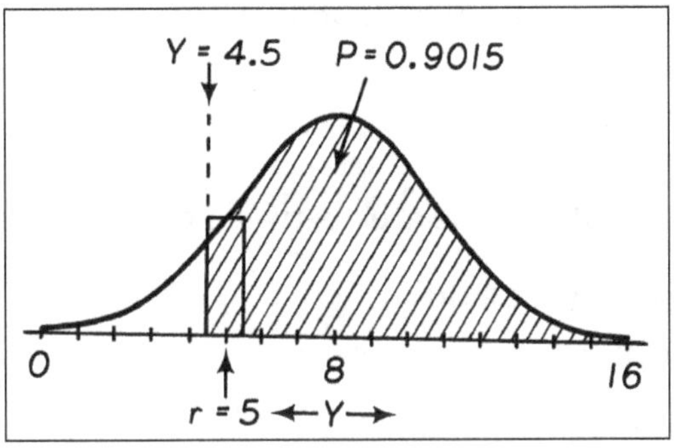

Step 2. Y = r - 1/2 = 5 - 1/2 = 4.5.

Step 3. Z = (4.5 - 8.0) / 2.71 = -1.29.

Step 4. $P\ (Z \leq -1.29)\ =\ P\ (Z \leq 1.29) = 0.0985$.

Step 5. The problem is to find the probability that $P\ (Z \leq -1.29)\ =\ P\ (Z \leq 1.29) = 0.9015$.

(d) To find the probability that more than 4 but less than 15 cans are underweight, we must find the probability of 5 more and 14 or less underweight cans as in Figure.

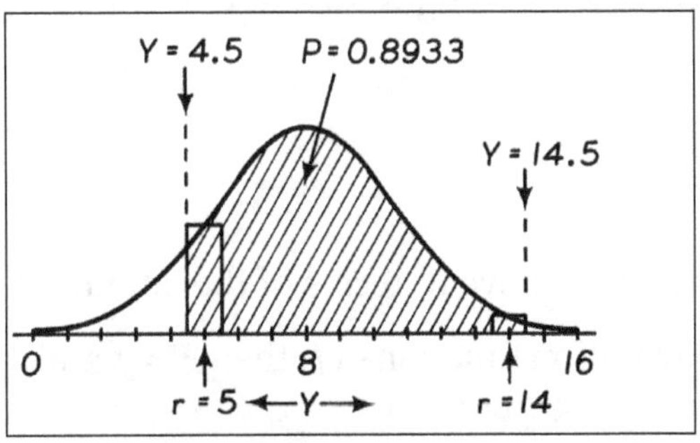

Step 2. Note in this case that we need to find the probability between two r values, r_1 and r_2.

$Y_1 = r_1 - 1/2 = 5 - 1/2 = 4.5$

$Y_2 = r_2 + 1/2 = 14 + 1/2 = 14.5$

Step 3. Now we have to calculate two standardized Z values, Z_1 and Z_2.

$$Z_1 = \frac{4.5-8.0}{2.71} \; 1.29$$

$$Z_2 = \frac{14.5-8.0}{2.71} \; 2.40$$

Step 4. The two probabilities:

$$P\,(Z \leq -1.29) \;=\; P\,(Z \leq 1.29) = 0.0985.$$

$$P(Z \geq 2.40) = 0.0082$$

Step 5. The required probability is the area between the two Z values, which is equal to:

$$\begin{aligned} &1 - P\,(Z \leq -1.29) - P(Z \geq 2.40) \\ &= 1 - 0.0985 - 0.0082 \\ &= 1 - 0.1067 = 0.8933 \end{aligned}$$

SAMPLE DISTRIBUTION OF THE MEDIAN

In addition to the smallest (Y_1) and largest (Y_n) order statistics, we are often interested in the sample median, $\tilde{X}$. For a sample of odd size, n = 2m + 1, the sample median is defined as Y_{m+1}. If n = 2m is even, the sample median is defined as $\frac{1}{2}(Y_m + Y_{m+1})$. We will prove a relation between the sample median and the population median $\tilde{\mu}$. By definition, $\tilde{\mu}$ satisfies,

$$\int_{-\infty}^{\tilde{\mu}} f(x)dx = \frac{1}{2}$$

It is convenient to re-write the above in terms of the cumulative distribution function. If F is the cumulative distribution function of f, then $F' = f$ and $\int_{-\infty}^{\tilde{\mu}} f(x)dx = \frac{1}{2}$ becomes,

$$F(\tilde{\mu}) = \frac{1}{2}$$

We are now ready to consider the distribution of the sample median.

Median Theorem: Let a sample of size n = 2m + 1 with n large be taken from an infinite population with a density function $f(\tilde{x})$ that is nonzero at the population median $\tilde{\mu}$

and continuously differentiable in a neighbourhood of $\tilde{\mu}$. The sampling distribution of the median is approximately normal with mean $\tilde{\mu}$ and variance $\frac{1}{8f(\tilde{\mu})^2 m}$.

Proof: Let the median random variable $\tilde{X}$ have values $\tilde{x}$ and density g($\tilde{x}$). The median is simply the (m + 1)th order statistic, so its distribution is given by the result. By Theorem,

$$g(\tilde{x}) = \frac{(2m+1)!}{m!m!}\left[\int_{-\infty}^{\tilde{x}} f(\tilde{x})dx\right]^m f(\tilde{x})\left[\int_{-\infty}^{\infty} f(x)dx\right]^m$$

We will first find an approximation for the constant factor in this equation. For this, we will use Stirling's approximation, which tells us that $n! = n^n e^{-n}\sqrt{2\pi n}\left(1 + O\left(n^{-1}\right)\right)$. We will consider values sufficiently large so that the terms of order 1/n need not be considered. Hence,

$$\frac{(2m+1)!}{m!m!} = \frac{(2m+1)(2m)!}{(m!)^2} \approx \frac{(2m+1)(2m)^{2m}e^{-2m}\sqrt{2\pi(2m)}}{\left(m^m e^{-m}\sqrt{2\pi m}\right)^2} = \frac{(2m+1)4^m}{\sqrt{\pi m}}$$

As F is the cumulative distribution function, $F(\bar{x}) = \int_{-\infty}^{\tilde{x}} f(x)dx$, which implies,

$$g(\tilde{x}) \approx \frac{\sim(2m+1)4^m}{\sqrt{\pi m}}\left[F(\tilde{x})\right]^m f(\tilde{x})\left[1-F(\tilde{x})\right]^m$$

We will need the Taylor series expansion of F($\tilde{x}$) about $\tilde{\mu}$, which is just,

$$F(\tilde{x}) = F(\tilde{\mu}) + F'(\tilde{\mu})(\tilde{x} - \tilde{\mu}) + O\left((\tilde{x}-\tilde{\mu})^2\right)$$

Because $\tilde{\mu}$ is the population median, $F(\tilde{\mu}) = 1/2$. Further, since F is the cumulative distribution function, F′ = f and we find,

$$F(\tilde{x}) = \frac{1}{2} + f(\tilde{\mu})(\tilde{x} - \tilde{\mu}) + O\left((\tilde{x} - \tilde{\mu})^2\right)$$

This approximation is only useful if $\tilde{x} - \tilde{\mu}$ is small; in other words, we need,

$$\lim_{m\to\infty} |\tilde{x} - \tilde{\mu}| = 0.$$

Letting t = $\tilde{x} - \tilde{\mu}$ (which is small and tends to 0 as m → ∞), substituting our Taylor series expansion into $g(\tilde{x}) \approx \frac{\sim(2m+1)4^m}{\sqrt{\pi m}}\left[F(\tilde{x})\right]^m f(\tilde{x})\left[1-F(\tilde{x})\right]^m$ yields,

$$g(\tilde{x}) \approx \frac{(2m+1)4^m}{\sqrt{\pi m}}\left[\frac{1}{2}+f(\tilde{\mu})t+O\left(t^2\right)\right]^m f(\tilde{x})\left[1-\left(\frac{1}{2}+f(\tilde{\mu})t+O\left(t^2\right)\right)\right]^m$$

By rearranging and combining factors, we find that,

$$g(\tilde{x}) \approx \frac{(2m+1)4^m}{\sqrt{\pi m}} f(\tilde{x})\left[\frac{1}{4}+f(\tilde{\mu})t+O(t^2)\right]^m f(\tilde{x})\left[1-\left(\frac{1}{2}+f(\tilde{\mu})t+O(t^2)\right)\right]^m$$

Remember that one definition of e^x is,

$$e^{x} = exp(x) = lim_{n\to\infty}\left(1+\frac{x}{n}\right)^n$$

Using this, and ignoring higher powers of t for the moment, we have for large m that,

$$g(\tilde{x}) \approx \frac{(2m+1)(\tilde{x})}{\sqrt{\pi m}}\exp\left(-4mf(\tilde{\mu})^2 t^2\right) \approx \frac{(2m+1)(\tilde{x})}{\sqrt{\pi m}}\exp\left(-\frac{(\tilde{x}-\tilde{\mu})^2}{1/\left(4mf(\tilde{\mu})^2\right)}\right)$$

$\tilde{x}$ can be assumed arbitrarily close to $\tilde{u}$ with high probability, we can assume $f(\tilde{x}) \approx f(\tilde{\mu})$ so that,

$$g(\tilde{x}) \approx \frac{(2m+1)f(\tilde{\mu})}{\sqrt{\pi m}}\exp\left(-\frac{(\tilde{x}-\tilde{\mu})^2}{1/\left(4mf(\tilde{\mu})^2\right)}\right)$$

Looking at the exponential part of the expression for $g(\tilde{x})$, we see that it appears to be a normal density with mean $\tilde{\mu}$ and $\sigma^2 = 1/(8mf(\tilde{\mu})^2)$. If we were instead to compute the variance from the normalization constant, we would find the variance to be,

$$\frac{m}{2(2m+1)2f(\tilde{\mu})^{2}}$$

We see that the two values are asymptotically equivalent, thus we can take the variance to be $\sigma^2 = 1/(8mf(\tilde{\mu})^2)$. Thus to complete the proof of the theorem, all that we need to is prove that we may ignore the higher powers of t and replace the product with an exponential in passing from previous equations. We have

$$\left(1-\frac{\left(4m\left(f(\tilde{\mu})t\right)^2\right)}{m}+O(t^3)\right)^m = \exp\left(m\log\left(1-4\left(f(\tilde{\mu})t\right)^2+O(t^3)\right)\right)$$

We use the Taylor series expansion of log(1 – x):

$$\log(1-x) = -x + O(x^2);$$

We only need one term in the expansion as t is small.

$$\left(1-\frac{4m\left(f(\tilde{\mu})t\right)^2}{m}+O\left(t^3\right)\right)^m = \exp\left(-m\cdot 4\left(f(\tilde{\mu})t\right)^2 + O\left(mt^3\right)\right)$$

$$= \exp\left(-\frac{(\tilde{x}-\tilde{\mu})^2}{1/\left(4mf(\tilde{\mu})^2\right)}\right)\cdot \exp\left(O\left(mt^3\right)\right).$$

One can show that as m → ∞, mt³ → 0. Thus the exp(O(mt³)) term above tends to 1, which completes the proof.

Remark. Our justification of ignoring the higher powers of t and replacing the product with an exponential in passing from previous equation is a standard technique. Namely, we replace some quantity (1 – P) m with (1 – P) m = exp(m log(1 – P)), Taylor expand the logarithm, and then look at the limit as m → ∞.

SAMPLING DISTRIBUTION OF STANDARD DEVIATION

Consider the sample standard deviation,

$$s \equiv \sqrt{\frac{1}{N}\sum_{i=1}^{N}(x_i-\bar{x})^2}$$

for n samples taken from a population with a normal distribution. The distribution of s is then given by,

$$f_N(s) = 2\frac{\left(\frac{N}{2\sigma^2}\right)^{(N-1)/2}}{\Gamma\left(\frac{1}{2}(N-1)\right)}e^{-Ns^2/\left(2\sigma^2\right)}s^{N-2},$$

where $\Gamma(z)$ is a gamma function and

$$\sigma^2 \equiv \frac{Ns^2}{N-1}$$

The function $f_N(s)$ is plotted above for $N=2$ (red), 4 (orange), ..., 10 (blue), and 12 (violet).

The mean is given by,

$$\langle s\rangle = \sqrt{\frac{2}{N}}\frac{\Gamma\left(\frac{N}{2}\right)}{\Gamma\left(\frac{N-1}{2}\right)}\sigma$$

$$\equiv b(N)\sigma,$$

where,

$$b(N) = \sqrt{\frac{2}{N}}\frac{\Gamma\left(\frac{N}{2}\right)}{\Gamma\left(\frac{N-1}{2}\right)}$$

The function $b(N)$ is known as $c4$ in statistical process control (Duncan 1986, pp. 62 and 134). Romanovsky showed that,

$$b(N)=1-\frac{3}{4N}-\frac{7}{32N^2}-\frac{9}{128N^3}+\ldots$$

The raw moments are given by,

$$\mu_r' = \left(\frac{2}{N}\right)^{r/2}\frac{\Gamma\left(\frac{N-1+r}{2}\right)}{\Gamma\left(\frac{N-1}{2}\right)}\sigma^2$$

and the variance of s is,

$$\operatorname{var}(s) = \mu_2' - \mu^2$$

$$= \frac{1}{N}\left[N-1-\frac{2\Gamma^2\left(\frac{N}{2}\right)}{\Gamma^2\left(\frac{N-1}{2}\right)}\right]\sigma^2.$$

$s/b(N)$ is an unbiased estimator of σ.

References

- Sampling-distribution, basic-statistics: emathzone.com, Retrieved 15 June, 2019
- Examples-of-sampling-distribution, basic-statistics: emathzone.com, Retrieved 27 March, 2019
- Important-sampling-distributions, research-methodology-tutorial-11512: wisdomjobs.com, Retrieved 07 May, 2019
- StandardDeviationDistribution: mathworld.wolfram.com, Retrieved 15 August, 2019
- An Invitation to Modern Number Theory, Princeton University Press, Princeton, NJ, 2006

5

Statistical Inference

Statistical inference is the process that makes use of data analysis for deducing properties of a probability distribution. Algorithmic inference, fiducial inference and Bayesian inference fall under its domain. This chapter closely examines the varied aspects of statistical inference to provide an extensive understanding of the subject.

Statstical Inference is the process of drawing conclusions about a parameter one is seeking to measure or estimate. Often scientists have many measurements of an object—say, the mass of an electron—and wish to choose the best measure. One principal approach of statistical inference is Bayesian estimation, which incorporates reasonable expectations or prior judgments (perhaps based on previous studies), as well as new observations or experimental results. Another method is the likelihood approach, in which "prior probabilities" are eschewed in favour of calculating a value of the parameter that would be most "likely" to produce the observed distribution of experimental outcomes.

In parametric inference, a particular mathematical form of the distribution function is assumed. Nonparametric inference avoids this assumption and is used to estimate parameter values of an unknown distribution having an unknown functional form.

Sampling in Statistical Inference

The use of randomization in sampling allows for the analysis of results using the methods of statistical inference. Statistical inference is based on the laws of probability, and allows analysts to infer conclusions about a given population based on results observed through random sampling. Two of the key terms in statistical inference are *parameter* and *statistic*:

A *parameter* is a number describing a population, such as a percentage or proportion.

A *statistic* is a number which may be computed from the data observed in a random sample without requiring the use of any unknown parameters, such as a sample mean.

Suppose an analyst wishes to determine the percentage of defective items which are produced by a factory over the course of a week. Since the factory produces thousands of items per week, the analyst takes a sample 300 items and observes that 15 of these are defective. Based on these results, the analyst computes the statistic $\hat{p}$, 15/300 = 0.05, as an estimate of the *parameter p*, or true proportion of defective items in the entire population.

Suppose the analyst takes 200 samples, of size 300 each, from the same group of items, and achieves the following results:

Number of Samples	Percentage of Defective Items
20	3
30	4
50	5
45	6
35	7
30	8

The histogram corresponding to these results is shown below:

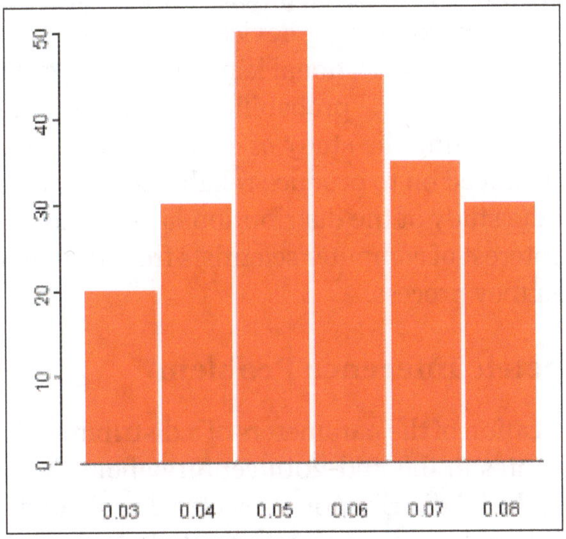

These results approximate a sampling distribution for the statistic $\hat{p}$, or the distribution of values taken by the statistic in all possible samples of the size 300 from the population of factory items. The distribution appears to be approximately normal, with mean between 0.05 and 0.06. With repeated sampling, the sampling distribution would more closely approximate a normal distribution, although it would remain discontinuous because of the granularity caused by rounding to percentage points.

Bias and Variability

When a statistic $\hat{p}$ is systematically skewed away from the true parameter p, it is considered to be a biased estimator of the parameter. In the factory example above, if the true percentage of defective items was known to be 8%, then our sampling distribution would be biased in the direction of estimating too few defective items. An unbiased estimator will have a sampling distribution whose mean is equal to the true value of the parameter.

The *variability* of a statistic is determined by the spread of its sampling distribution. In general, larger samples will have smaller variability. This is because as the sample size increases, the chance of observing extreme values decreases and the observed values for the statistic will group more closely around the mean of the sampling distribution. Furthermore, if the population size is significantly larger than the sample size, then the size of the population will not affect the variability of the sampling distribution (i.e., a sample of size 100 from a population of size 100,000 will have the same variability as a sample of size 100 from a population of size 1,000,000).

ALGORITHMIC INFERENCE

Algorithmic inference gathers new developments in the statistical inference methods made feasible by the powerful computing devices widely available to any data analyst. Cornerstones in this field are computational learning theory, granular computing, bioinformatics, and, long ago, structural probability. The main focus is on the algorithms which compute statistics rooting the study of a random phenomenon, along with the amount of data they must feed on to produce reliable results. This shifts the interest of mathematicians from the study of the distribution laws to the functional properties of the statistics, and the interest of computer scientists from the algorithms for processing data to the information they process.

The Fisher Parametric Inference Problem

Concerning the identification of the parameters of a distribution law, the mature reader may recall lengthy disputes in the mid-20th century about the interpretation of their variability in terms of fiducial distribution , structural probabilities, priors/posteriors, and so on. From an epistemology viewpoint, this entailed a companion dispute as to the nature of probability: is it a physical feature of phenomena to be described through random variables or a way of synthesizing data about a phenomenon? Opting for the latter, Fisher defines a *fiducial distribution* law of parameters of a given random variable that he deduces from a sample of its specifications. With this law he computes, for instance "the probability that μ (mean of a Gaussian variable is less than any assigned value, or the probability that it lies between any assigned values, or, in short, its probability distribution, in the light of the sample observed".

The Classic Solution

Fisher fought hard to defend the difference and superiority of his notion of parameter distribution in comparison to analogous notions, such as Bayes' posterior distribution,

Fraser's constructive probability and Neyman's confidence intervals. For half a century, Neyman's confidence intervals won out for all practical purposes, crediting the phenomenological nature of probability. With this perspective, when you deal with a Gaussian variable, its mean μ is fixed by the physical features of the phenomenon you are observing, where the observations are random operators, hence the observed values are specifications of a random sample. Because of their randomness, you may compute from the sample specific intervals containing the fixed μ with a given probability that you denote *confidence*.

Example:

Let X be a Gaussian variable with parameters μ and σ^2 and $\{X_1,\ldots,X_m\}$ a sample drawn from it. Working with statistics,

$$S_\mu = \sum_{i=1}^{m} X_i$$

and

$$S_{\sigma^2} = \sum_{i=1}^{m} (X_i - \bar{X})^2, \text{ where } \bar{X} = \frac{S_\mu}{m}$$

is the sample mean, we recognize that,

$$T = \frac{S_\mu - m\mu}{\sqrt{S_{\sigma^2}}}\sqrt{\frac{m-1}{m}} = \frac{\bar{X}-\mu}{\sqrt{S_{\sigma^2}/(m(m-1))}}$$

follows a Student's t distribution with parameter (degrees of freedom) $m - 1$, so that,

$$f_T(t) = \frac{\Gamma(m/2)}{\Gamma((m-1)/2)}\frac{1}{\sqrt{\pi(m-1)}}\left(1+\frac{t^2}{m-1}\right)^{m/2}.$$

Gauging T between two quantiles and inverting its expression as a function of μ you obtain confidence intervals for μ.

With the sample specification:

$$\mathbf{x} = \{7.14, 6.3, 3.9, 6.46, 0.2, 2.94, 4.14, 4.69, 6.02, 1.58\}$$

having size $m = 10$, you compute the statistics $s_\mu = 43.37$ and $s_{\sigma^2} = 46.07$, , and obtain a 0.90 confidence interval for μ with extremes (3.03, 5.65).

Inferring Functions with the Help of a Computer

From a modeling perspective the entire dispute looks like a chicken-egg dilemma: Either fixed data by first and probability distribution of their properties as a consequence, or fixed properties by first and probability distribution of the observed data as a corollary. The classic solution has one benefit and one drawback. The former was appreciated particularly back when people still did computations with sheet and pencil. Per se, the task of computing a Neyman confidence interval for the fixed parameter θ is hard: you don't know θ, but you look for disposing around it an interval with a possibly very low probability of failing. The analytical solution is allowed for a very limited number of theoretical cases. *Vice versa* a large variety of instances may be quickly solved in an *approximate way* via the central limit theorem in terms of confidence interval around a Gaussian distribution – that's the benefit. The drawback is that the central limit theorem is applicable when the sample size is sufficiently large. Therefore, it is less and less applicable with the sample involved in modern inference instances. The fault is not in the sample size on its own part. Rather, this size is not sufficiently large because of the complexity of the inference problem.

With the availability of large computing facilities, scientists refocused from isolated parameters inference to complex functions inference, i.e. re sets of highly nested parameters identifying functions. In these cases we speak about *learning of functions* (in terms for instance of regression, neuro-fuzzy system or computational learning) on the basis of highly informative samples. A first effect of having a complex structure linking data is the reduction of the number of sample degrees of freedom, i.e. the burning of a part of sample points, so that the effective sample size to be considered in the central limit theorem is too small. Focusing on the sample size ensuring a limited learning error with a given confidence level, the consequence is that the lower bound on this size grows with complexity indices such as VC dimension or detail of a class to which the function we want to learn belongs.

A sample of 1,000 independent bits is enough to ensure an absolute error of at most 0.081 on the estimation of the parameter p of the underlying Bernoulli variable with a confidence of at least 0.99. The same size cannot guarantee a threshold less than 0.088 with the same confidence 0.99 when the error is identified with the probability that a 20-year-old man living in New York does not fit the ranges of height, weight and waistline observed on 1,000 Big Apple inhabitants. The accuracy shortage occurs because both the VC dimension and the detail of the class of parallelepipeds, among which the one observed from the 1,000 inhabitants' ranges falls, are equal to 6.

General Inversion Problem Solving the Fisher Question

With insufficiently large samples, the approach: *fixed sample – random properties* suggests inference procedures in three steps:

- Sampling mechanism: It consists of a pair (Z, g_θ), where the seed Z is a random variable without unknown parameters, while the explaining function g_θ is a function mapping from samples of Z to samples of the random variable X

we are interested in. The parameter vector θ is a specification of the random parameter Θ. Its components are the parameters of the X distribution law. The Integral Transform Theorem ensures the existence of such a mechanism for each (scalar or vector) X when the seed coincides with the random variable U uniformly distributed in $[0,1]$.

Example: For X following a Pareto distribution with parameters a and k, i.e,

$$F_X(x) = \left(1 - \frac{k^a}{x}\right) I_{[k,\infty)}(x),$$

a sampling mechanism $(U, g_{(a,k)})$ for X with seed U reads:

$$g_{(a,k)}(u) = k(1-u)^{-\frac{1}{a}},$$

or, equivalently, $g_{(a,k)}(u) = ku^{-1/a}$.

- Master equations: The actual connection between the model and the observed data is tossed in terms of a set of relations between statistics on the data and unknown parameters that come as a corollary of the sampling mechanisms. We call these relations master equations. Pivoting around the statistic.

$s = h(x_1, \ldots, x_m) = h(g_\theta(z_1), \ldots, g_\theta(z_m))$, the general form of a master equation is:

$$s = \rho(\theta; z_1, \ldots, z_m).$$

With these relations we may inspect the values of the parameters that could have generated a sample with the observed statistic from a particular setting of the seeds representing the seed of the sample. Hence, to the population of sample seeds corresponds a population of parameters. In order to ensure this population clean properties, it is enough to draw randomly the seed values and involve either sufficient statistics or, simply, well-behaved statistics w.r.t. the parameters, in the master equations.

For example, the statistics $s_1 = \sum_{i=1}^{m} \log x_i$ and $s_2 = \min_{i=1,\ldots,m} \{x_i\}$ prove to be sufficient for parameters a and k of a Pareto random variable X. Thanks to the (equivalent form of the) sampling mechanism $g_{(a,k)}$ we may read them as,

$$s_1 = m \log k + 1/a \sum_{i=1}^{m} \log u_i$$

$$s_2 = \min_{i=1,\ldots,m} \{k u_i^{-\frac{1}{a}}\},$$

respectively.

- Parameter population: Having fixed a set of master equations, you may map sample seeds into parameters either numerically through a population bootstrap, or analytically through a twisting argument. Hence from a population of seeds you obtain a population of parameters.

Example: From the above master equation we can draw a pair of parameters, compatible with the observed sample by solving the following system of equations,

$$a = \frac{\sum \log u_i - m \log \min\{u_i\}}{s_1 - m \log s_2}.$$

$$k = e^{\frac{as_1 - \sum \log u_i}{ma}}$$

where s_1 and s_2 are the observed statistics and $u_1,\ldots,u_m$ a set of uniform seeds. Transferring to the parameters the probability (density) affecting the seeds, you obtain the distribution law of the random parameters A and K compatible with the statistics you have observed.

Compatibility denotes parameters of compatible populations, i.e. of populations that *could have generated* a sample giving rise to the observed statistics. You may formalize this notion as follows:

For a random variable and a sample drawn from it a *compatible distribution* is a distribution having the same sampling mechanism $\mathcal{M}_X = (Z, g_\theta)$ of X with a value θ of the random parameter Θ derived from a master equation rooted on a well-behaved statistic s.

Example:

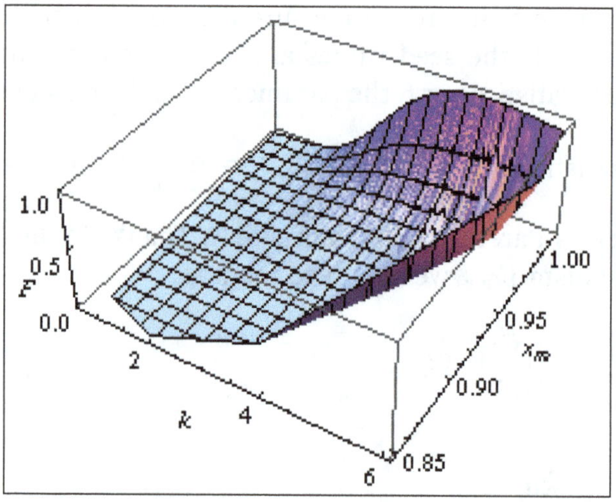

Joint empirical cumulative distribution function of parameters (A, K) of a Pareto random variable.

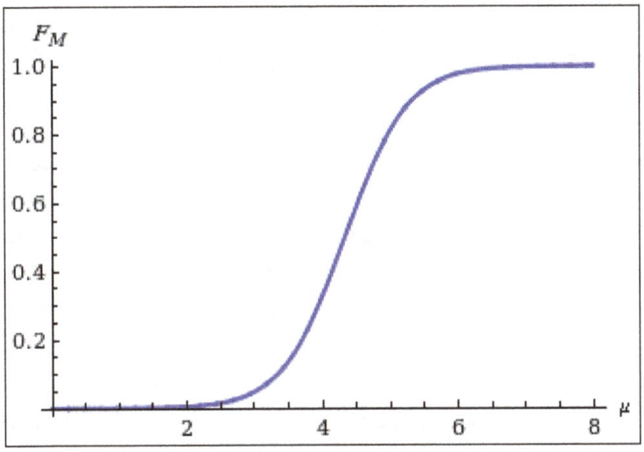

Cumulative distribution function of the mean M of a Gaussian random variable.

You may find the distribution law of the Pareto parameters A and K as an implementation example of the population bootstrap method as in the figure on the left.

Implementing the twisting argument method, you get the distribution law $F_M(\mu)$ of the mean M of a Gaussian variable X on the basis of the statistic $s_M = \sum_{i=1}^{m} x_i$ when Σ^2 is known to be equal to σ^2. Its expression is:

$$F_M(\mu) = \Phi\left(\frac{m\mu - s_M}{\sigma\sqrt{m}}\right),$$

shown in the figure on the right, where Φ is the cumulative distribution function of a standard normal distribution.

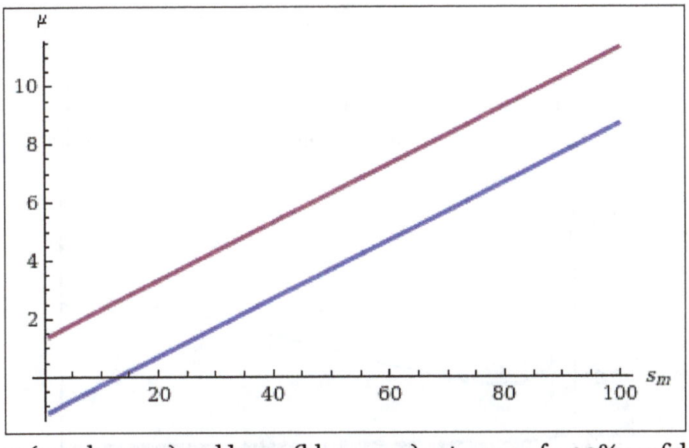

Upper (purple curve) and lower (blue curve) extremes of a 90% confidence interval of the mean M of a Gaussian random variable for a fixed σ and different values of the statistic s_m.

Computing a confidence interval for M given its distribution function is straightforward: we need only find two quantiles (for instance $\delta/2$ and $1-\delta/2$ quantiles in case we are interested in a confidence interval of level δ symmetric in the tail's probabilities) as indicated on the left in the diagram showing the behavior of the two bounds for different values of the statistic s_m.

The Achilles heel of Fisher's approach lies in the joint distribution of more than one parameter, say mean and variance of a Gaussian distribution. On the contrary, with the last approach (and above-mentioned methods: population bootstrap and twisting argument) we may learn the joint distribution of many parameters. For instance, focusing on the distribution of two or many more parameters, in the figures below we report two confidence regions where the function to be learnt falls with a confidence of 90%. The former concerns the probability with which an extended support vector machine attributes a binary label 1 to the points of the (x, y) plane. The two surfaces are drawn on the basis of a set of sample points in turn labelled according to a specific distribution law. The latter concerns the confidence region of the hazard rate of breast cancer recurrence computed from a censored sample.

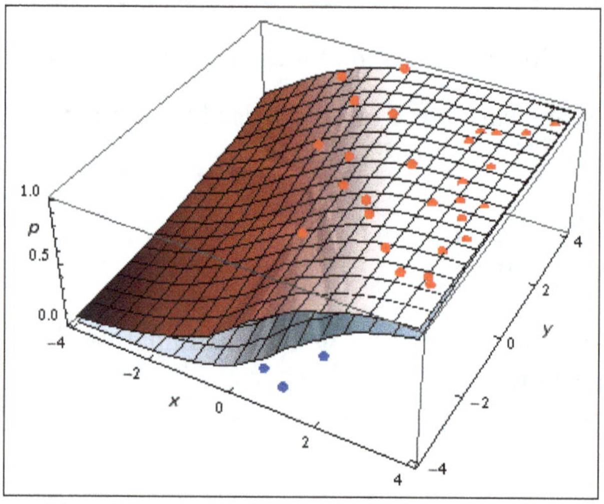

90% confidence region for the family of support vector machines endowed with hyperbolic tangent profile function.

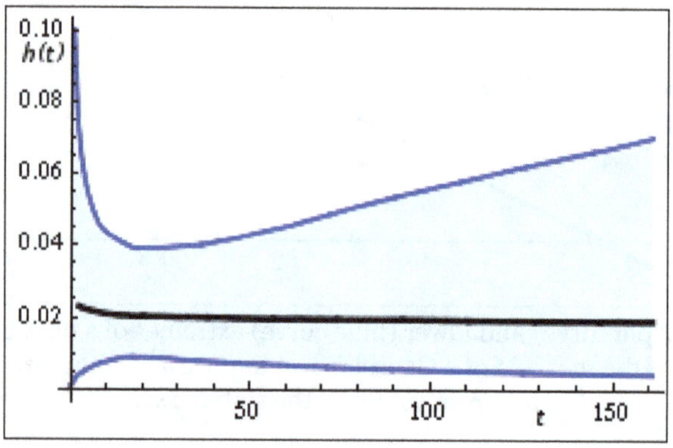

90% confidence region for the hazard function of breast cancer recurrence computed from the censored sample $t = (9,13,>13,18,12,23,31,34,>45,48,>161)$ with > t denoting a censored time.

Twisting Properties

Starting with a sample $\{x_1,\ldots,x_m\}$ observed from a random variable X having a given distribution law with a non-set parameter, a parametric inference problem consists of computing suitable values – call them estimates – of this parameter precisely on the basis of the sample. An estimate is suitable if replacing it with the unknown parameter does not cause major damage in next computations. In algorithmic inference, suitability of an estimate reads in terms of compatibility with the observed sample.

In turn, parameter compatibility is a probability measure that we derive from the probability distribution of the random variable to which the parameter refers. In this way we identify a random parameter Θ compatible with an observed sample. Given a sampling mechanism $M_X = (g_\theta, Z)$, the rationale of this operation lies in using the Z seed distribution law to determine both the X distribution law for the given θ, and the Θ distribution law given an X sample. Hence, we may derive the latter distribution directly from the former if we are able to relate domains of the sample space to subsets of Θ support. In more abstract terms, we speak about twisting properties of samples with properties of parameters and identify the former with statistics that are suitable for this exchange, so denoting a well behavior w.r.t. the unknown parameters. The operational goal is to write the analytic expression of the cumulative distribution function $F_\Theta(\theta)$, , in light of the observed value s of a statistic S, as a function of the S distribution law when the X parameter is exactly θ.

Method

Given a sampling mechanism $M_X = (g_\theta, Z)$ for the random variable X, we model $\mathbf{X} = \{X_1,\ldots,X_m\}$ to be equal to $\{g_\theta(Z_1),\ldots,g_\theta(Z_m)\}$. Focusing on a relevant statistic $S = h_1(X_1,\ldots,X_m)$ for the parameter θ, the master equation reads,

$$s = h(g_\theta(z_1),\ldots,g_\theta(z_m)) = \rho(\theta; z_1,\ldots,z_m).$$

When s is a well-behaved statistic w.r.t the parameter, we are sure that a monotone relation exists for each $z = \{z_1,\ldots,z_m\}$ between s and θ. We are also assured that Θ, as a function of Z for given s, is a random variable since the master equation provides solutions that are feasible and independent of other (hidden) parameters.

The direction of the monotony determines for any z a relation between events of the type $s \geq s' \leftrightarrow \theta \geq \theta'$ or *vice versa* $s \geq s' \leftrightarrow \theta \leq \theta'$, where s' is computed by the master equation with θ'. In the case that s assumes discrete values the first relation changes into $s \geq s' \rightarrow \theta \geq \theta' \rightarrow s \geq s' + \ell$ where $\ell > 0$ is the size of the s discretization grain, idem with the opposite monotony trend. Resuming these relations on all seeds, for s continuous we have either,

$$F_{\Theta|S=s}(\theta) = F_{S|\Theta=\theta}(s)$$

or

$$F_{\Theta|S=s}(\theta) = 1 - F_{S|\Theta=\theta}(s)$$

For s discrete we have an interval where $F_{\Theta|S=s}(\theta)$ lies, because of $\ell > 0$. The whole logical contrivance is called a twisting argument. A procedure implementing it is as follows.

Algorithm

Generating a Parameter Distribution law through a Twisting Argument

Given a sample $\{x_1,\ldots,x_m\}$ from a random variable with parameter θ unknown,

- Identify a well behaving statistic S for the parameter θ and its discretization grain ℓ (if any).
- Decide the monotony versus.
- Compute $F_\Theta(\theta) \in (q_1(F_{S|\Theta=\theta}(s)), q_2(F_{S|\Theta=\theta}(s)))$ where:
    - if S is continuous $q_1 = q_2$.
    - if S is discrete.
        - $q_2(F_S(s)) = q_1(F_S(s-\ell)$ if s does not decrease with θ.
        - $q_1(F_S(s)) = q_2(F_S(s-\ell)$ if s does not increase with θ.
        - $q_i(F_S) = 1 - F_S$ if s does not decrease with θ and $q_i(F_S) = F_S$ if s does not increase with θ for $i = 1,2$.

The rationale behind twisting arguments does not change when parameters are vectors, though some complication arises from the management of joint inequalities. Instead, the difficulty of dealing with a vector of parameters proved to be the Achilles heel of Fisher's approach to the fiducial distribution of parameters. Also Fraser's constructive probabilities devised for the same purpose do not treat this point completely.

Example:

For $\boldsymbol{x}$ drawn from a gamma distribution, whose specification requires values for the parameters λ and k, a twisting argument may be stated by following the below procedure. Given the meaning of these parameters we know that,

$$(k \le k') \leftrightarrow (s_k \le s_{k'}) \text{ for fixed } \lambda,$$

$$(\lambda \le \lambda') \leftrightarrow (s_{\lambda'} \le s_\lambda) \text{ for fixed } k,$$

where $s_k = \prod_{i=1}^{m} x_i$ and $s_\lambda = \sum_{i=1}^{m} x_i$. This leads to a joint cumulative distribution function

$$F_{\Lambda,K}(\lambda,k) = F_{\Lambda|K=k}(\lambda)F_K(k) = F_{K|\Lambda=\lambda}(k)F_\Lambda(\lambda).$$

Using the first factorization and replacing s_k with $r_k = \frac{s_k}{s_\lambda^m}$ in order to have a distribution of K that is independent of Λ, we have,

$$F_{\Lambda|K=k}(\lambda) = 1 - \frac{\Gamma(km, \lambda s_\Lambda)}{\Gamma(km)}$$

$$F_K(k) = 1 - F_{R_k}(r_K)$$

with m denoting the sample size, s_Λ and r_K are the observed statistics (hence with indices denoted by capital letters), $\tilde{A}(a,b)$ the incomplete gamma function and $F_{R_k}(r_K)$ the Fox's H function that can be approximated with a gamma distribution again with proper parameters (for instance estimated through the method of moments) as a function of k and m.

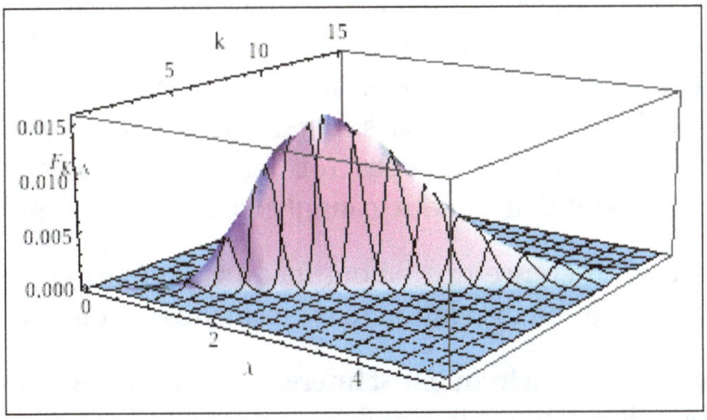

Joint probability density function of parameters (K,Λ) of a Gamma random variable.

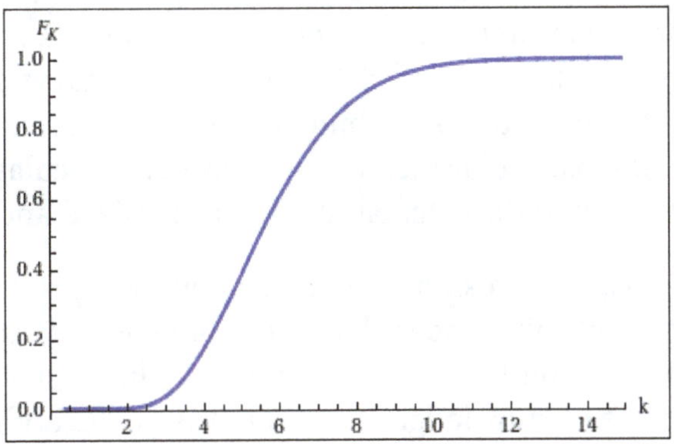

Marginal cumulative distribution function of parameter K of a Gamma random variable.

With a sample size $m = 30, s_{\Lambda} = 72.82$ and $r_K = 4.5 \times 10^{-46}$, you may find the joint p.d.f. of the Gamma parameters K and Λ on the left. The marginal distribution of K is reported in the picture on the right.

FIDUCIAL INFERENCE

Fiducial inference is one of a number of different types of statistical inference. These are rules, intended for general application, by which conclusions can be drawn from samples of data. In modern statistical practice, attempts to work with fiducial inference have fallen out of fashion in favour of frequentist inference, Bayesian inference and decision theory. However, fiducial inference is important in the history of statistics since its development led to the parallel development of concepts and tools in theoretical statistics that are widely used. Some current research in statistical methodology is either explicitly linked to fiducial inference or is closely connected to it.

The general approach of fiducial inference was proposed by Ronald Fisher. Here "fiducial" comes from the Latin for faith. Fiducial inference can be interpreted as an attempt to perform inverse probability without calling on prior probability distributions. Fiducial inference quickly attracted controversy and was never widely accepted. Indeed, counter-examples to the claims of Fisher for fiducial inference were soon published. These counter-examples cast doubt on the coherence of "fiducial inference" as a system of statistical inference or inductive logic. Other studies showed that, where the steps of fiducial inference are said to lead to "fiducial probabilities" (or "fiducial distributions"), these probabilities lack the property of additivity, and so cannot constitute a probability measure.

The concept of fiducial inference can be outlined by comparing its treatment of the problem of interval estimation in relation to other modes of statistical inference.

- A confidence interval, in frequentist inference, with coverage probability γ has the interpretation that among all confidence intervals computed by the same method, a proportion γ will contain the true value that needs to be estimated. This has either a repeated sampling (or frequentist) interpretation, or is the probability that an interval calculated from yet-to-be-sampled data will cover the true value. However, in either case, the probability concerned is not the probability that the true value is in the particular interval that has been calculated since at that stage both the true value and the calculated interval are fixed and are not random.
- Credible intervals, in Bayesian inference, do allow a probability to be given for the event that an interval, once it has been calculated does include the true value, since it proceeds on the basis that a probability distribution can be associated with the state of knowledge about the true value, both before and after the sample of data has been obtained.

Fisher designed the fiducial method to meet perceived problems with the Bayesian approach, at a time when the frequentist approach had yet to be fully developed. Such problems related to the need to assign a prior distribution to the unknown values. The aim was to have a procedure, like the Bayesian method, whose results could still be given an inverse probability interpretation based on the actual data observed. The method proceeds by attempting to derive a "fiducial distribution", which is a measure of the degree of faith that can be put on any given value of the unknown parameter and is faithful to the data in the sense that the method uses all available information.

Unfortunately Fisher did not give a general definition of the fiducial method and he denied that the method could always be applied. His only examples were for a single parameter; different generalisations have been given when there are several parameters. A relatively complete presentation of the fiducial approach to inference is given by Quenouille, while Williams describes the application of fiducial analysis to the calibration problem (also known as "inverse regression") in regression analysis.

The Fiducial Distribution

Fisher required the existence of a sufficient statistic for the fiducial method to apply. Suppose there is a single sufficient statistic for a single parameter. That is, suppose that the conditional distribution of the data given the statistic does not depend on the value of the parameter. For example, suppose that n independent observations are uniformly distributed on the interval $[0,\omega]$. The maximum, X, of the n observations is a sufficient statistic for ω. If only X is recorded and the values of the remaining observations are forgotten, these remaining observations are equally likely to have had any values in the interval $[0,X]$. This statement does not depend on the value of ω. Then X contains all the available information about ω and the other observations could have given no further information.

The cumulative distribution function of X is,

$$F(x)=P(X\leq x)=P(\text{all observations}\leq x)=\left(\frac{x}{\omega}\right)^n.$$

Probability statements about X/ω may be made. For example, given α, a value of a can be chosen with $0 < a < 1$ such that,

$$P(X>\omega a)=1-a^n=\alpha.$$

Thus,

$$a=(1-\alpha)^{\frac{1}{n}}.$$

Then Fisher might say that this statement may be inverted into the form,

$$P\left(\omega < \frac{X}{a}\right) = \alpha.$$

In this latter statement, ω is now regarded as variable and X is fixed, whereas previously it was the other way round. This distribution of ω is the *fiducial distribution* which may be used to form fiducial intervals that represent degrees of belief.

The calculation is identical to the pivotal method for finding a confidence interval, but the interpretation is different. In fact older books use the terms *confidence interval* and *fiducial interval* interchangeably. Notice that the fiducial distribution is uniquely defined when a single sufficient statistic exists.

The pivotal method is based on a random variable that is a function of both the observations and the parameters but whose distribution does not depend on the parameter. Such random variables are called pivotal quantities. By using these, probability statements about the observations and parameters may be made in which the probabilities do not depend on the parameters and these may be inverted by solving for the parameters in much the same way as in the example above. However, this is only equivalent to the fiducial method if the pivotal quantity is uniquely defined based on a sufficient statistic.

A fiducial interval could be taken to be just a different name for a confidence interval and give it the fiducial interpretation. But the definition might not then be unique. Fisher would have denied that this interpretation is correct: for him, the fiducial distribution had to be defined uniquely and it had to use all the information in the sample.

Status of the Approach

Fisher admitted that "fiducial inference" had problems. Fisher wrote to George A. Barnard that he was "not clear in the head" about one problem on fiducial inference, and, also writing to Barnard, Fisher complained that his theory seemed to have only "an asymptotic approach to intelligibility". Later Fisher confessed that "I don't understand yet what fiducial probability does. We shall have to live with it a long time before we know what it's doing for us. But it should not be ignored just because we don't yet have a clear interpretation".

Lindley showed that fiducial probability lacked additivity, and so was not a probability measure. Cox points out that the same argument applies to the so-called "confidence distribution" associated with confidence intervals, so the conclusion to be drawn from this is moot. Fisher sketched "proofs" of results using fiducial probability. When the conclusions of Fisher's fiducial arguments are not false, many have been shown to also follow from Bayesian inference.

In 1978, J. G. Pederson wrote that "the fiducial argument has had very limited success and is now essentially dead". Davison wrote "A few subsequent attempts have been made to resurrect fiducialism, but it now seems largely of historical importance, particularly in view of its restricted range of applicability when set alongside models of current interest."

However, fiducial inference is still being studied and its principles appear valuable for some scientific applications. In the mid-2010s, the psychometrician Yang Liu developed generalized fiducial inference for models in item response theory and demonstrated favorable results compared to frequentist and Bayesian approaches. Other current work in fiducial inference is ongoing under the name of confidence distributions.

BAYESIAN INFERENCE

Bayesian inference is a method of statistical inference in which Bayes' theorem is used to update the probability for a hypothesis as more evidence or information becomes available. Bayesian inference is an important technique in statistics, and especially in mathematical statistics. Bayesian updating is particularly important in the dynamic analysis of a sequence of data. Bayesian inference has found application in a wide range of activities, including science, engineering, philosophy, medicine, sport, and law. In the philosophy of decision theory, Bayesian inference is closely related to subjective probability, often called "Bayesian probability".

Introduction to Bayes' Rule

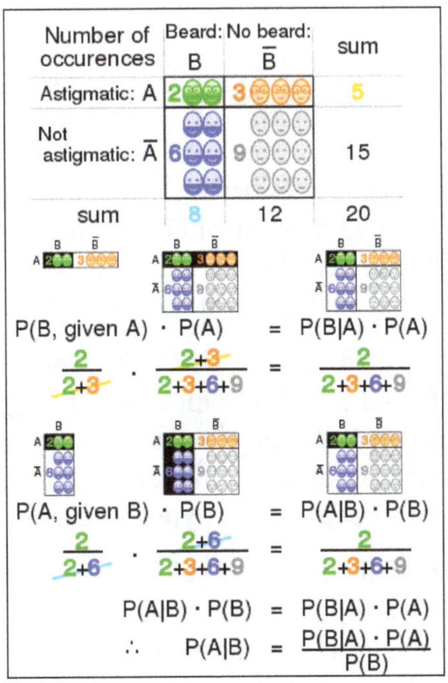

A geometric visualisation of Bayes' theorem. In the table, the values 2, 3, 6 and 9 give the relative weights of each corresponding condition and case. The figures denote the cells of the table involved in each metric, the probability being the fraction of each figure that is shaded. This shows that P(A|B) P(B) = P(B|A) P(A) i.e. $P(A|B) = \frac{P(B|A)\,P(A)}{P(B)}$. Similar reasoning can be used to show that $P(\bar{A}|B) = \frac{P(B|\bar{A})\,P(\bar{A})}{P(B)}$ etc.

Formal Explanation

Bayesian inference derives the posterior probability as a consequence of two antecedents: a prior probability and a "likelihood function" derived from a statistical model for the observed data. Bayesian inference computes the posterior probability according to Bayes' theorem:

$$P(H \mid E) = \frac{P(E \mid H) \cdot P(H)}{P(E)}$$

where,

- H stands for any hypothesis whose probability may be affected by data (called evidence below). Often there are competing hypotheses, and the task is to determine which is the most probable.
- $P(H)$, the prior probability, is the estimate of the probability of the hypothesis H before the data E, the current evidence, is observed.
- E, the evidence, corresponds to new data that were not used in computing the prior probability.
- $P(H \mid E)$, the posterior probability, is the probability of H given E, i.e., after E is observed. This is what we want to know: the probability of a hypothesis given the observed evidence.
- $P(E \mid H)$ is the probability of observing E given H, and is called the likelihood. As a function of E with H fixed, it indicates the compatibility of the evidence with the given hypothesis. The likelihood function is a function of the evidence, H, while the posterior probability is a function of the hypothesis, H.
- $P(E)$ is sometimes termed the marginal likelihood or "model evidence". This factor is the same for all possible hypotheses being considered (as is evident from the fact that the hypothesis H does not appear anywhere in the symbol, unlike for all the other factors), so this factor does not enter into determining the relative probabilities of different hypotheses.

For different values of H, only the factors $P(H)$ and $P(E \mid H)$, both in the numerator, affect the value of $P(H \mid E)$ – the posterior probability of a hypothesis is proportional to its prior probability (its inherent likeliness) and the newly acquired likelihood (its compatibility with the new observed evidence).

Bayes' rule can also be written as follows:

$$P(H \mid E) = \frac{P(E \mid H)}{P(E)} \cdot P(H)$$

where the factor $\frac{P(E \mid H)}{P(E)}$ can be interpreted as the impact of E on the probability of H.

Probability of a Hypothesis

Suppose there are two full bowls of cookies. Bowl #1 has 10 chocolate chip and 30 plain cookies, while bowl #2 has 20 of each. Our friend Fred picks a bowl at random, and then picks a cookie at random. We may assume there is no reason to believe Fred treats one bowl differently from another, likewise for the cookies. The cookie turns out to be a plain one. How probable is it that Fred picked it out of bowl #1?

Intuitively, it seems clear that the answer should be more than a half, since there are more plain cookies in bowl #1. The precise answer is given by Bayes' theorem. Let H_1 correspond to bowl #1, and H_2 to bowl #2. It is given that the bowls are identical from Fred's point of view, thus $P(H_1) = P(H_2)$, and the two must add up to 1, so both are equal to 0.5. The event E is the observation of a plain cookie. From the contents of the bowls, we know that $P(E \mid H_1) = 30/40 = 0.75$ and $P(E \mid H_2) = 20/40 = 0.5$. Bayes' formula then yields,

$$\begin{aligned} P(H_1 \mid E) &= \frac{P(E \mid H_1)P(H_1)}{P(E \mid H_1)P(H_1) + P(E \mid H_2)P(H_2)} \\ &= \frac{0.75 \times 0.5}{0.75 \times 0.5 + 0.5 \times 0.5} \\ &= 0.6 \end{aligned}$$

Before we observed the cookie, the probability we assigned for Fred having chosen bowl #1 was the prior probability, $P(H_1)$, which was 0.5. After observing the cookie, we must revise the probability to $P(H_1 \mid E)$, which is 0.6.

Making a Prediction

An archaeologist is working at a site thought to be from the medieval period, between the 11th century to the 16th century. However, it is uncertain exactly when in this period the site was inhabited. Fragments of pottery are found, some of which are glazed and some of which are decorated. It is expected that if the site were inhabited during the early medieval period, then 1% of the pottery would be glazed and 50% of its area decorated, whereas if it had been inhabited in the late medieval period then 81% would be glazed and 5% of its area decorated. How confident can the archaeologist be in the date of inhabitation as fragments are unearthed?

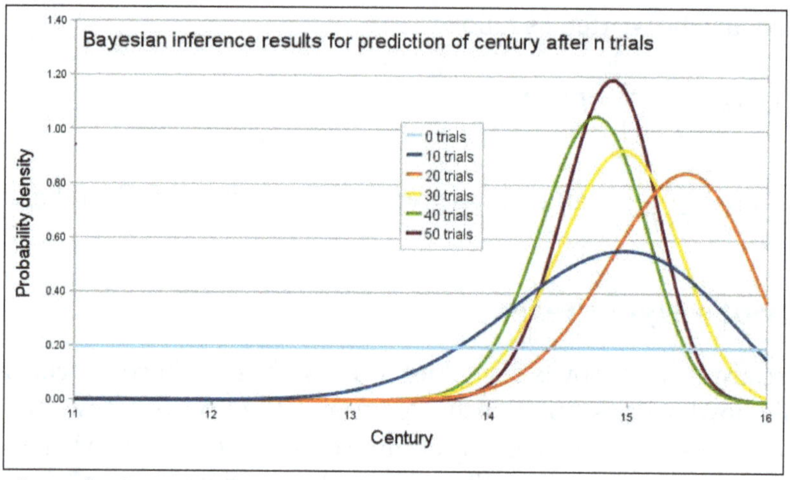

Example results for archaeology example. This simulation was generated using c=15.2.

The degree of belief in the continuous variable C (century) is to be calculated, with the discrete set of events $\{GD, G\bar{D}, \bar{G}D, \bar{G}\bar{D}\}$ as evidence. Assuming linear variation of glaze and decoration with time, and that these variables are independent,

$$P(E = GD \mid C = c) = (0.01 + \frac{0.81 - 0.01}{16 - 11}(c - 11))(0.5 - \frac{0.5 - 0.05}{16 - 11}(c - 11))$$

$$P(E = G\bar{D} \mid C = c) = (0.01 + \frac{0.81 - 0.01}{16 - 11}(c - 11))(0.5 + \frac{0.5 - 0.05}{16 - 11}(c - 11))$$

$$P(E = \bar{G}D \mid C = c) = ((1 - 0.01) - \frac{0.81 - 0.01}{16 - 11}(c - 11))(0.5 - \frac{0.5 - 0.05}{16 - 11}(c - 11))$$

$$P(E = \bar{G}\bar{D} \mid C = c) = ((1 - 0.01) - \frac{0.81 - 0.01}{16 - 11}(c - 11))(0.5 + \frac{0.5 - 0.05}{16 - 11}(c - 11))$$

Assume a uniform prior of $f_C(c) = 0.2$, and that trials are independent and identically distributed. When a new fragment of type e is discovered, Bayes' theorem is applied to update the degree of belief for each c:

$$f_C(c \mid E = e) = \frac{P(E = e \mid C = c)}{P(E = e)} f_C(c) = \frac{P(E = e \mid C = c)}{\int_{11}^{16} P(E = e \mid C = c) f_C(c) dc} f_C(c)$$

A computer simulation of the changing belief as 50 fragments are unearthed is shown on the graph. In the simulation, the site was inhabited around 1420, or $c = 15.2$. By calculating the area under the relevant portion of the graph for 50 trials, the archaeologist can say that there is practically no chance the site was inhabited in the 11th and 12th centuries, about 1% chance that it was inhabited during the 13th century, 63% chance during the 14th century and 36% during the 15th century. The Bernstein-von Mises theorem asserts here the asymptotic convergence to the "true" distribution because the probability space corresponding to the discrete set of events $\{GD, G\bar{D}, \bar{G}D, \bar{G}\bar{D}\}$ is finite.

In Frequentist Statistics and Decision Theory

A decision-theoretic justification of the use of Bayesian inference was given by Abraham Wald, who proved that every unique Bayesian procedure is admissible. Conversely, every admissible statistical procedure is either a Bayesian procedure or a limit of Bayesian procedures.

Wald characterized admissible procedures as Bayesian procedures (and limits of Bayesian procedures), making the Bayesian formalism a central technique in such areas of frequentist inference as parameter estimation, hypothesis testing, and computing confidence intervals. For example:

- "Under some conditions, all admissible procedures are either Bayes procedures or limits of Bayes procedures (in various senses). These remarkable results, at least in their original form, are due essentially to Wald. They are useful because the property of being Bayes is easier to analyze than admissibility".
- "In decision theory, a quite general method for proving admissibility consists in exhibiting a procedure as a unique Bayes solution".
- "In the first chapters of this work, prior distributions with finite support and the corresponding Bayes procedures were used to establish some of the main theorems relating to the comparison of experiments. Bayes procedures with respect to more general prior distributions have played a very important role in the development of statistics, including its asymptotic theory." "There are many problems where a glance at posterior distributions, for suitable priors, yields immediately interesting information. Also, this technique can hardly be avoided in sequential analysis".
- "A useful fact is that any Bayes decision rule obtained by taking a proper prior over the whole parameter space must be admissible".
- "An important area of investigation in the development of admissibility ideas has been that of conventional sampling-theory procedures, and many interesting results have been obtained".

Probabilistic Programming

While conceptually simple, Bayesian methods can be mathematically and numerically challenging. Probabilistic programming languages (PPLs) implement functions to easily build Bayesian models together with efficient automatic inference methods. This helps separate the model building from the inference, allowing practitioners to focus on their specific problems and leaving PPLs to handle the computational details for them.

Applications

Computer Applications

Bayesian inference has applications in artificial intelligence and expert systems. Bayesian inference techniques have been a fundamental part of computerized pattern recognition techniques since the late 1950s. There is also an ever-growing connection between Bayesian methods and simulation-based Monte Carlo techniques since complex models cannot be processed in closed form by a Bayesian analysis, while a graphical model structure may allow for efficient simulation algorithms like the Gibbs sampling and other Metropolis–Hastings algorithm schemes. Recently Bayesian inference has gained popularity among the phylogenetics community for these reasons; a number of applications allow many demographic and evolutionary parameters to be estimated simultaneously.

As applied to statistical classification, Bayesian inference has been used in recent years to develop algorithms for identifying e-mail spam. Applications which make use of Bayesian inference for spam filtering include CRM114, DSPAM, Bogofilter, SpamAssassin, SpamBayes, Mozilla, XEAMS, and others.

Solomonoff's Inductive inference is the theory of prediction based on observations; for example, predicting the next symbol based upon a given series of symbols. The only assumption is that the environment follows some unknown but computable probability distribution. It is a formal inductive framework that combines two well-studied principles of inductive inference: Bayesian statistics and Occam's Razor. Solomonoff's universal prior probability of any prefix p of a computable sequence x is the sum of the probabilities of all programs (for a universal computer) that compute something starting with p. Given some p and any computable but unknown probability distribution from which x is sampled, the universal prior and Bayes' theorem can be used to predict the yet unseen parts of x in optimal fashion.

Bioinformatic and Healthcare Applications

Bayesian inference has been applied in different Bioinformatics applications, including differential gene expression analysis. Bayesian inference is also used in a general cancer risk model, called CIRI (Continuous Individualized Risk Index), where serial measurements are incorporated to update a Bayesian model which is primarily built from prior knowledge.

In the Courtroom

Bayesian inference can be used by jurors to coherently accumulate the evidence for and against a defendant, and to see whether, in totality, it meets their personal threshold for 'beyond a reasonable doubt'. Bayes' theorem is applied successively to all evidence presented, with the posterior from one stage becoming the prior for the next. The benefit of a Bayesian approach is that it gives the juror an unbiased, rational mechanism for combining evidence. It may be appropriate to explain Bayes' theorem to jurors in odds form, as betting odds are more widely understood than probabilities. Alternatively, a logarithmic approach, replacing multiplication with addition, might be easier for a jury to handle.

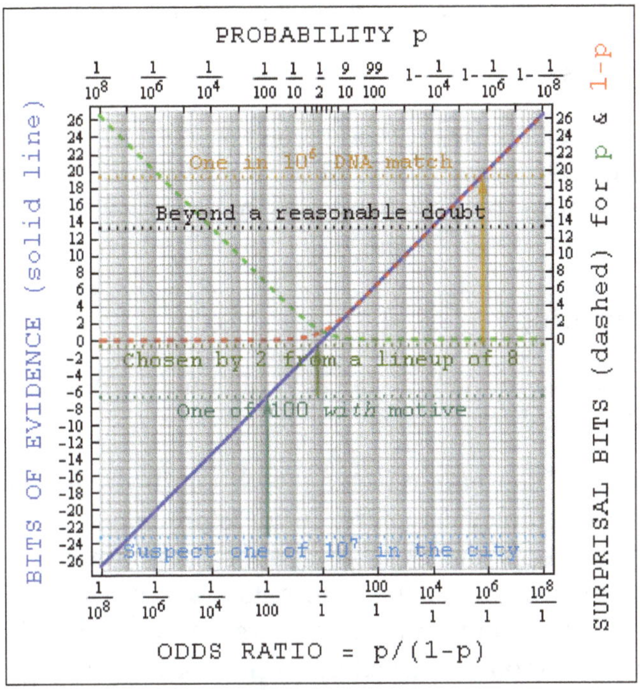

Adding up evidence.

If the existence of the crime is not in doubt, only the identity of the culprit, it has been suggested that the prior should be uniform over the qualifying population. For example, if 1,000 people could have committed the crime, the prior probability of guilt would be 1/1000.

The use of Bayes' theorem by jurors is controversial. In the United Kingdom, a defence expert witness explained Bayes' theorem to the jury in *R v Adams*. The jury convicted, but the case went to appeal on the basis that no means of accumulating evidence had been provided for jurors who did not wish to use Bayes' theorem. The Court of Appeal upheld the conviction, but it also gave the opinion that "To introduce Bayes' Theorem, or any similar method, into a criminal trial plunges the jury into inappropriate and unnecessary realms of theory and complexity, deflecting them from their proper task".

Gardner-Medwin argues that the criterion on which a verdict in a criminal trial should be based is *not* the probability of guilt, but rather the *probability of the evidence, given that the defendant is innocent* (akin to a frequentist p-value). He argues that if the posterior probability of guilt is to be computed by Bayes' theorem, the prior probability of guilt must be known. This will depend on the incidence of the crime, which is an unusual piece of evidence to consider in a criminal trial. Consider the following three propositions:

- The known facts and testimony could have arisen if the defendant is guilty.
- The known facts and testimony could have arisen if the defendant is innocent.
- The defendant is guilty.

Gardner-Medwin argues that the jury should believe both A and not-B in order to convict. A and not-B implies the truth of C, but the reverse is not true. It is possible that B and C are both true, but in this case he argues that a jury should acquit, even though they know that they will be letting some guilty people go free.

Bayesian Epistemology

Bayesian epistemology is a movement that advocates for Bayesian inference as a means of justifying the rules of inductive logic.

Karl Popper and David Miller have rejected the idea of Bayesian rationalism, i.e. using Bayes rule to make epistemological inferences: It is prone to the same vicious circle as any other justificationist epistemology, because it presupposes what it attempts to justify. According to this view, a rational interpretation of Bayesian inference would see it merely as a probabilistic version of falsification, rejecting the belief, commonly held by Bayesians, that high likelihood achieved by a series of Bayesian updates would prove the hypothesis beyond any reasonable doubt, or even with likelihood greater than 0.

Other

- The scientific method is sometimes interpreted as an application of Bayesian inference. In this view, Bayes' rule guides (or should guide) the updating of probabilities about hypotheses conditional on new observations or experiments. The Bayesian inference has also been applied to treat stochastic scheduling problems with incomplete information by Cai.
- Bayesian search theory is used to search for lost objects.
- Bayesian inference in phylogeny.
- Bayesian tool for methylation analysis.
- Bayesian approaches to brain function investigate the brain as a Bayesian mechanism.
- Bayesian inference in ecological studies.
- Bayesian inference is used to estimate parameters in stochastic chemical kinetic models.

Bayesian Inference in Marketing

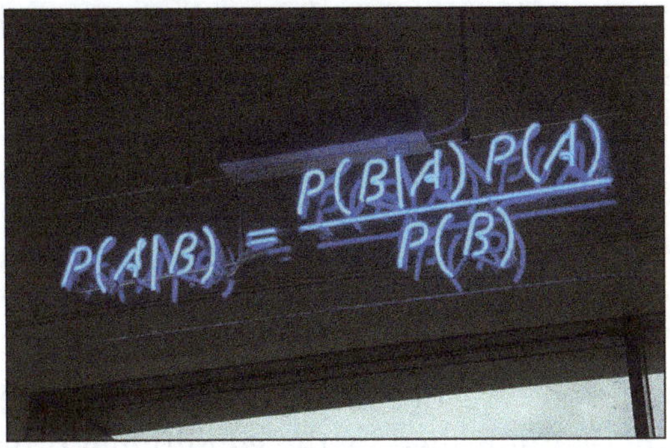

Bayes' Theorem.

In marketing, Bayesian inference allows for decision making and market research evaluation under uncertainty and with limited data.

Bayes' theorem is fundamental to Bayesian inference. It is a subset of statistics, providing a mathematical framework for forming inferences through the concept of probability, in which evidence about the true state of the world is expressed in terms of degrees of belief through subjectively assessed numerical probabilities. Such a probability is known as a Bayesian probability. The fundamental ideas and concepts behind Bayes' theorem, and its use within Bayesian inference, have been developed and added to over the past centuries by Thomas Bayes, Richard Price and Pierre Simon Laplace as well as numerous other mathematicians, statisticians and scientists. Bayesian inference has experienced spikes in popularity as it has been seen as vague and controversial by rival frequentist statisticians. In the past few decades Bayesian inference has become widespread in many scientific and social science fields such as marketing. Bayesian inference allows for decision making and market research evaluation under uncertainty and limited data.

Bayes' Theorem

Bayesian probability specifies that there is some prior probability. Bayesian statisticians can use both an objective and a subjective approach when interpreting the prior probability, which is then updated in light of new relevant information. The concept is a manipulation of conditional probabilities:

$$P(AB) = P(A \mid B)P(B) = P(B \mid A)P(A)$$

Alternatively, a more simple understanding of the formula may be reached by substituting the events A and B to become respectively the hypothesis (H) and the data (D). The rule allows for a judgment of the relative truth of the hypothesis given the data.

This is done through the calculation shown below, where $P(D \mid H)$ is the likelihood function. This assesses the probability of the observed data (D) arising from the hypothesis (H); $P(H)$ is the assigned prior probability or initial belief about the hypothesis;

the denominator $P(D)$ is formed by the integrating or summing of $P(D|H)P(H)$; $P(H \mid D)$ is known as the posterior which is the recalculated probability, or updated belief about the hypothesis. It is a result of the prior beliefs as well as sample information. The posterior is a conditional distribution as the result of collecting or in consideration of new relevant data.

$$P(H \mid D) = \frac{P(D|H)P(H)}{P(D)}$$

To sum up this formula: the posterior probability of the hypothesis is equal to the prior probability of the hypothesis multiplied by the conditional probability of the evidence given the hypothesis, divided by the probability of the new evidence.

Use in Marketing

Bayesian decision theory can be applied to all four areas of the marketing mix. Assessments are made by a decision maker on the probabilities of events that determine the profitability of alternative actions where the outcomes are uncertain. Assessments are also made for the profit (utility) for each possible combination of action and event. The decision maker can decide how much research, if any, needs to be conducted in order to investigate the consequences associated with the courses of action under evaluation. This is done before a final decision is made, but in order to do this costs would be incurred, time used and may overall be unreliable. For each possible action, expected profit can be computed, that is a weighted mean of the possible profits, the weights being the probabilities. The decision maker can then choose the action for which the expected profit is the highest. The theorem provides a formal reconciliation between judgment expressed quantitatively in the prior distribution and the statistical evidence of the experiment.

New Product Development

The use of Bayesian decision theory in new product development allows for the use of subjective prior information. Bayes in new product development allows for the comparison of additional review project costs with the value of additional information in order to reduce the costs of uncertainty. The methodology used for this analysis is in the form of decision trees and 'stop'/'go' procedures. If the predicted payoff (the posterior) is acceptable for the organisation the project should go ahead, if not, development should stop. By reviewing the posterior (which then becomes the new prior) on regular intervals throughout the development stage managers are able to make the best possible decision with the information available at hand.

Pricing Decisions

Bayesian decision theory can be used in looking at pricing decisions. Field information such as retail and wholesale prices as well as the size of the market and market share are all incorporated into the prior information. Managerial judgement is included in order to evaluate different pricing strategies. This method of evaluating possible pricing strategies does have its limitations as it requires a number of assumptions to be made about the market place in which an organisation operates. As markets are dynamic environments it is often difficult to fully apply Bayesian decision theory to pricing strategies without simplifying the model.

Promotional Campaigns

When dealing with promotion a marketing manager must account for all the market complexities that are involved in a decision. As it is difficult to account for all aspects of the market, a manager should look to incorporate both experienced judgements from senior executives as well modifying these judgements in light of economically justifiable information gathering. An example of the application of Bayesian decision theory for promotional purposes could be the use of a test sample in order to assess the effectiveness of a promotion prior to a full scale rollout. By combining prior subjective data about the occurrence of possible events with experimental empirical evidence gained through a test market, the resultant data can be used to make decisions under risk.

Channel Decisions and the Logistics of Distribution

Bayesian decision analysis can also be applied to the channel selection process. In order to help provide further information the method can be used that produces results in a profit or loss aspect. Prior information can include costs, expected profit, training expenses and any other costs relevant to the decision as well as managerial experience which can be displayed in a normal distribution. Bayesian decision making under uncertainty lets a marketing manager assess his/her options for channel logistics by computing the most profitable method choice. A number of different costs can be entered into the model that helps to assess the ramifications of change in distribution method. Identifying and quantifying all of the relevant information for this process can be very time consuming and costly if the analysis delays possible future earnings.

Strengths

The Bayesian approach is superior to use in decision making when there is a high level of uncertainty or limited information in which to base decisions on and where expert opinion or historical knowledge is available. Bayes is also useful when explaining the findings in a probability sense to people who are less familiar and comfortable with comprehending statistics. It is in this sense that Bayesian methods are thought of as having created a bridge between business judgments and statistics for the purpose of decision-making.

The three principle strengths of Bayes' theorem that have been identified by scholars are that it is prescriptive, complete and coherent. Prescriptive in that it is the theorem that is the simple prescription to the conclusions reached on the basis of evidence and reasoning for the consistent decision maker. It is complete because the solution is often clear and unambiguous, for a given choice of model and prior distribution. It allows for the incorporation of prior information when available to increase the robustness of the solutions, as well as taking into consideration the costs and risks that are associated with choosing alternative decisions. Lastly Bayes theorem is coherent. It is considered the most appropriate way to update beliefs by welcoming the incorporation of new information, as is seen through the probability distributions. This is further complemented by the fact that Bayes inference satisfies the likelihood principle, which states that models or inferences for datasets leading to the same likelihood function should generate the same statistical information. Bayes methods are more cost effective than the traditional frequentist take on marketing research and subsequent decision making. The probability can be assessed from a degree of belief before and after accounting for evidence, instead of calculating the probabilities of a certain decision by carrying out a large number of trials with each one producing an outcome from a set of possible outcomes. The planning and implementation of trials to see how a decision impacts in the 'field' e.g. observing consumers reaction to a relabeling of a product, is time consuming and costly, a method many firms cannot afford. In place of taking the frequentist route in aiming for a universally acceptable conclusion through iteration, it is sometimes more effective to take advantage of all the information available to the firm to work out the 'best' decision at the time, and then subsequently when new knowledge is obtained, revise the posterior distribution to be then used as the prior, thus the inferences continue to logically contribute to one another based on Bayes theorem.

Weaknesses

In marketing situations, it is important that the prior probability is (1) chosen correctly, and (2) is understood. A disadvantage to using Bayesian analysis is that there is no 'correct' way to choose a prior, therefore the inferences require a thorough analysis to translate the subjective prior beliefs into a mathematically formulated prior to ensure that the results will not be misleading and consequently lead to the disproportionate analysis of preposteriors. The subjective definition of probability and the selection and use of the priors have led to statisticians critiquing this subjective definition of probability that underlies the Bayesian approach. Bayesian probability is often found to be difficult when analysing and assessing probabilities due to its initial counter intuitive nature. Often when deciding between strategies based on a decision, they are interpreted as: where there is evidence X that shows condition A might hold true, is misread by judging A's likelihood by how well the evidence X matches A, but crucially without considering the prior frequency of A. In alignment with Falsification, which aims to question and falsify instead of prove hypotheses, where there is very strong evidence X, it does not necessarily mean there is a very high probability that A leads to B, but in fact should be interpreted as a very low probability of A not leading to B. In the field of marketing, behavioural experiments which have dealt with managerial decision-making, and risk perception, in consumer decisions have utilised the Bayesian model, or similar models, but found that it may not be relevant quantitatively in predicting human information processing behaviour. Instead the model has been proven as useful as a qualitative means of describing how individuals combine new evidence with their predetermined judgements. Therefore, "the model may have some value as a first approximation to the development of descriptive choice theory" in consumer and managerial instances.

An advertising manager is deciding whether or not to increase the advertising for a product in a particular market. The Bayes approach to this decision suggests: 1) These alternative courses of action for which the consequences are uncertain are a necessary condition in order to apply Bayes'; 2) The advertising manager will pick the course of action which allows him to achieve some objective i.e. a maximum return on his advertising investment in the form of profit; 3) He must determine the possible consequences of each action into some measure of success (or loss) with which a certain objective is achieved.

This 3 component example explains how the payoffs are conditional upon which outcomes occur. The advertising manager can characterize the outcomes based on past experience and knowledge and devise some possible events that are more likely to occur than others. He can then assign to these events prior probabilities, which would be in the form of numerical weights.

He can test out his predictions (prior probabilities) through an experiment. For example, he can run a test campaign to decide if the total level of advertising should be in fact increased. Based on the outcome of the experiment he can re-evaluate his prior probability and make a decision on whether to go ahead with increasing the advertising in the market or not. However gathering this additional data is costly, time consuming and may not lead to perfectly reliable results. As a decision makers he has to deal with experimental and systematic error and this is where Bayes' comes in.

It approaches the experimental problem by asking; is additional data required? If so, how much needs to be collected and by what means and finally, how does the decision maker revise his prior judgment in light of the results of the new experimental evidence? In this example the advertising manager can use the Bayesian approach to deal with his dilemma and update his prior judgments in light of new information he gains. He needs to take into account the profit (utility) attached to the alternative acts under different events and the value versus cost of information in order to make his optimal decision on how to proceed.

Bayes in Computational Models

Markov chain Monte Carlo (MCMC) is a flexible procedure designed to fit a variety of Bayesian models. It is the underlying method used in computational software such as the LaplacesDemon R Package and WinBUGS. The advancements and developments of these types of statistical software have allowed for the growth of Bayes by offering ease of calculation. This is achieved by the generation of samples from the posterior distributions, which are then used to produce a range of options or strategies which are allocated numerical weights. MCMC obtains these samples and produces summary and diagnostic statistics while also saving the posterior samples in the output. The decision maker can then assess the results from the output data set and choose the best option to proceed.

Bayesian Tool for Methylation Analysis

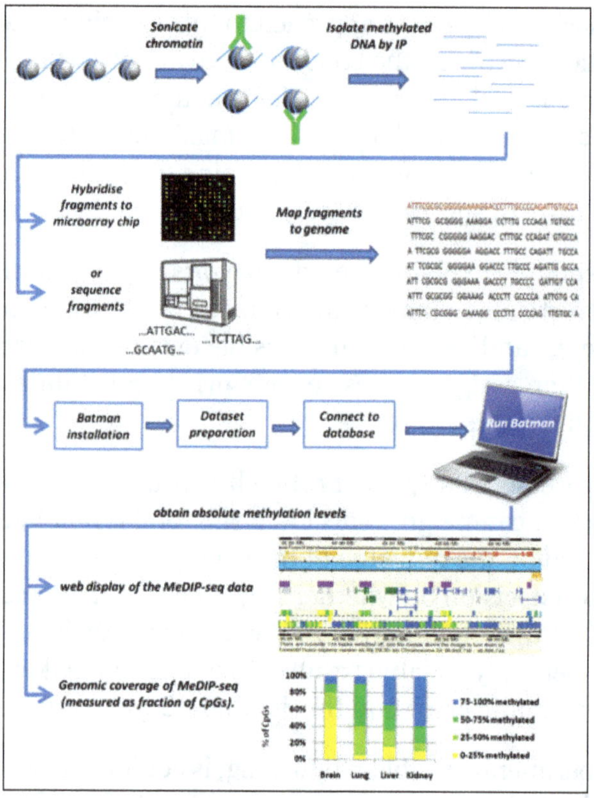

Batman workflow.

Bayesian tool for methylation analysis, also known as BATMAN, is a statistical tool for analysing methylated DNA immunoprecipitation (MeDIP) profiles. It can be applied to large datasets generated using either oligonucleotide arrays (MeDIP-chip) or next-generation sequencing (MeDIP-seq), providing a quantitative estimation of absolute methylation state in a region of interest.

MeDIP (methylated DNA immunoprecipitation) is an experimental technique used to assess DNA methylation levels by using an antibody to isolate methylated DNA sequences. The isolated fragments of DNA are either hybridized to a microarray chip (MeDIP-chip) or sequenced by next-generation sequencing (MeDIP-seq). While this tells you what areas of the genome are methylated, it does not give absolute methylation levels. Imagine two different genomic regions, *A* and *B*. Region *A* has six CpGs

(DNA methylation in mammalian somatic cells generally occurs at CpG dinucleotides), three of which are methylated. Region *B* has three CpGs, all of which are methylated. As the antibody simply recognizes methylated DNA, it will bind both these regions equally and subsequent steps will therefore show equal signals for these two regions. This does not give the full picture of methylation in these two regions (in region *A* only half the CpGs are methylated, whereas in region *B* all the CpGs are methylated). Therefore, to get the full picture of methylation for a given region you have to normalize the signal you get from the MeDIP experiment to the number of CpGs in the region, and this is what the Batman algorithm does. Analysing the MeDIP signal of the above example would give Batman scores of 0.5 for region *A* (i.e. the region is 50% methylated) and 1 for region *B* (i.e. The region is 100% methylated). In this way Batman converts the signals from MeDIP experiments to absolute methylation levels.

Development of Batman

The core principle of the Batman algorithm is to model the effects of varying density of CpG dinucleotides, and the effect this has on MeDIP enrichment of DNA fragments. The basic assumptions of Batman:

- Almost all DNA methylation in mammals happens at CpG dinucleotides.
- Most CpG-poor regions are constitutively methylated while most CpG-rich regions (CpG islands) are constitutively unmethylated.
- There are no fragment biases in MeDIP experiment (approximate range of DNA fragment sizes is 400–700 bp).
- The errors on the microarray are normally distributed with precision.
- Only methylated CpGs contribute to the observed signal.
- CpG methylation state is generally highly correlated over hundreds of bases, so CpGs grouped together in 50- or 100-bp windows would have the same methylation state.

Basic parameters in Batman:

- C_{cp}: Coupling factor between probe p and CpG dinucleotide *c*, is defined as the fraction of DNA molecules hybridizing to probe *p* that contain the CpG *c*.
- C_{tot}: Total CpG influence parameter, is defined as the sum of coupling factors for any given probe, which provides a measure of local CpG density.
- m_c: The methylation status at position *c*, which represents the fraction of chromosomes in the sample on which it is methylated. m_c is considered as a continuous variable since the majority samples used in MeDIP studies contain multiple cell-types.

Based on these assumptions, the signal from the MeDIP channel of the MeDIP-chip or MeDIP-seq experiment depends on the degree of enrichment of DNA fragments overlapping that probe, which in turn depends on the amount of antibody binding, and thus to the number of methylated CpGs on those fragments. In Batman model, the complete dataset from a MeDIP/chip experiment, A, can be represented by a statistical model in the form of the following probability distribution:

$$f(A \mid m) = \prod_{p} \phi\left(A_p \mid A_{base} + r\sum_{c} C_{cp}, v^{-1} \right),$$

where $\phi\ (x|\mu,\ \sigma^2)$ is a Gaussian probability density function. Standard Bayesian techniques can be used to infer $f(m|A)$, that is, the distribution of likely methylation states given one or more sets of MeDIP-chip/MeDIP-seq outputs. To solve this inference problem, Batman uses nested sampling to generate 100 independent samples from $f(m|A)$ for each tiled region of the genome, then summarizes the most likely methylation state in 100-bp windows by fitting beta distributions to these samples. The modes of the most likely beta distributions were used as final methylation calls.

Limitations

It may be useful to take the following points into account when considering using Batman:

- Batman is not a piece of software; it is an algorithm performed using the command prompt. As such it is not especially user-friendly and is quite a computationally technical process.
- Because it is non-commercial, there is very little support when using Batman beyond what is in the manual.
- It is quite time consuming (it can take several days to analyse one chromosome). (In one government lab, running Batman on a set of 100 Agilent Human DNA Methylation Arrays (about 250,000 probes per array) took less than an hour to complete in Agilent's Genomic Workbench software. Our computer had a 2GHz processor, 24 GB RAM, 64-bit Windows 7.)
- Copy number variation (CNV) has to be accounted for. For example, the score for a region with a CNV value of 1.6 in a cancer (a loss of 0.4 compared to normal) would have to be multiplied by 1.25 (=2/1.6) to compensate for the loss.
- One of the basic assumptions of Batman is that all DNA methylation occurs at CpG dinucleotides. While this is generally the case for vertebrate somatic cells, there are situations where there is widespread non-CpG methylation, such as in plant cells and embryonic stem cells.

Bayesian Inference in Phylogeny

Bayesian inference of phylogeny uses a likelihood function to create a quantity called the posterior probability of trees using a model of evolution, based on some prior probabilities, producing the most likely phylogenetic tree for the given data. The Bayesian approach has become popular due to advances in computing speeds and the integration of Markov chain Monte Carlo (MCMC) algorithms. Bayesian inference has a number of applications in molecular phylogenetics and systematics.

Bayesian Inference of Phylogeny Background and Bases

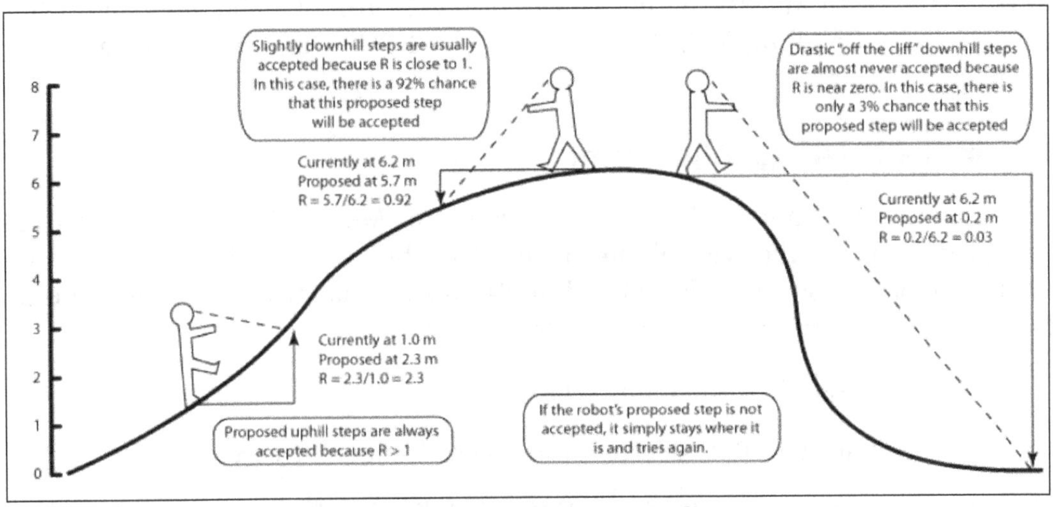

Metaphor illustrating MCMC method steps.

Bayesian inference refers to a probabilistic method developed by Reverend Thomas Bayes based on Bayes' theorem. Published posthumously in 1763 it was the first expression of inverse probability and the basis of Bayesian inference. Independently, unaware of Bayes work, Pierre-Simon Laplace developed Bayes' theorem in 1774.

Bayesian inference was widely used until 1900s when there was a shift to frequentist inference, mainly due to computational limitations. Based on Bayes' theorem, the bayesian approach combines the prior probability of a tree P(A) with the likelihood of the data (B) to produce a posterior probability distribution on trees P(A|B). The posterior probability of a tree will indicate the probability of the tree to be correct, being the tree with the highest posterior probability the one chosen to represent best a phylogeny. It was the introduction of Markov Chain Monte Carlo (MCMC) methods by Nicolas Metropolis in 1953 that revolutionized Bayesian Inference and by the 1990s became a widely used method amongst phylogeneticists. Some of the advantages over traditional maximum parsimony and maximum likelihood methods are the possibility of account for the phylogenetic uncertainty, use of prior information and incorporation of complex models of evolution that limited computational analyses for traditional methods. Although overcoming complex analytical operations the posterior probability still involves a summation over all trees and, for each tree, integration over all possible combinations of substitution model parameter values and branch lengths.

MCMC methods can be described in three steps: first using a stochastic mechanism a new state for the Markov chain is proposed. Secondly, the probability of this new state to be correct is calculated. Thirdly, a new random variable (0,1) is proposed. If this new value is less than the acceptance probability the new state is accepted and the state of the chain is updated. This process is run for either thousands or millions of times. The amount of time a single tree is visited during the course of the chain is just a valid approximation of its posterior probability. Some of the most common algorithms used in MCMC methods include the Metropolis-Hastings algorithms, the Metropolis-Coupling MCMC (MC^3) and the LOCAL algorithm of Larget and Simon.

Metropolis-Hastings Algorithm

One of the most common MCMC methods used is the Metropolis-Hastings algorithm, a modified version of the original Metropolis algorithm. It is a widely used method to sample randomly from complicated and multi-dimensional distribution probabilities. The Metropolis algorithm is described in the following steps:

- An initial tree, T_i, is randomly selected.
- A neighbour tree, T_j, is selected from the collection of trees.
- The ratio, R, of the probabilities (or probability density functions) of T_j and T_i is computed as follows: $R = f(T_j)/f(T_i)$.
- If $R \geq 1$, T_j is accepted as the current tree.
- If $R < 1$, T_j is accepted as the current tree with probability R, otherwise T_i is kept.
- At this point the process is repeated from Step 2 N times.

The algorithm keeps running until it reaches an equilibrium distribution. It also assumes that the probability of proposing a new tree T_j when we are at the old tree state T_i, is the same probability of proposing T_i when we are at T_j. When this is not the case Hastings corrections are applied. The aim of Metropolis-Hastings algorithm is to produce a collection of states with a determined distribution until the Markov process reaches a stationary distribution. The algorithm has two components:

- A potential transition from one state to another ($i \rightarrow j$) using a transition probability function $q_{i,j}$.
- Movement of the chain to state j with probability $\alpha_{i,j}$ and remains in i with probability $1 - \alpha_{i,j}$.

Metropolis-coupled MCMC

Metropolis-coupled MCMC algorithm (MC3) has been proposed to solve a practical concern of the Markov chain moving across peaks when the target distribution has multiple local peaks, separated by low valleys, are known to exist in the tree space. This is the case during heuristic tree search under maximum parsimony (MP), maximum likelihood (ML), and minimum evolution (ME) criteria, and the same can be expected for stochastic tree search using MCMC. This problem will result in samples not approximating correctly to the posterior density. The (MC3) improves the mixing of Markov chains in presence of multiple local peaks in the posterior density. It runs multiple (m) chains in parallel, each for n iterations and with different stationary distributions $\pi_j(.)$, $j=1,2,\ldots,m$,, where the first one, $\pi_1=\pi$ is the target density, while π_j , $j=2,3,\ldots,m$ are chosen to improve mixing. For example, one can choose incremental heating of the form:

$$\pi_j(\theta)=\pi(\theta)^{1/[1+\lambda(j-1)]},\ \lambda>0,$$

so that the first chain is the cold chain with the correct target density, while chains $2,3,\ldots,m$ are heated chains. Note that raising the density $\pi(.)$ to the power $1/T$ with $T>1$ has the effect of flattening out the distribution, similar to heating a metal. In such a distribution, it is easier to traverse between peaks (separated by valleys) than in the original distribution. After each iteration, a swap of states between two randomly chosen chains is proposed through a Metropolis-type step. Let $\theta^{(j)}$ be the current state in chain j, $j=1,2,\ldots,m$. A swap between the states of chains i and j is accepted with probability:

$$\alpha=\frac{\pi_i(\theta^{(j)})\pi_j(\theta^{(i)})}{\pi_i(\theta^{(i)})\pi_j(\theta^{(j)})}$$

At the end of the run, output from only the cold chain is used, while those from the hot chains are discarded. Heuristically, the hot chains will visit the local peaks rather easily, and swapping states between chains will let the cold chain occasionally jump valleys, leading to better mixing. However, if $\pi_i(\theta)/\pi_j(\theta)$ is unstable, proposed swaps will seldom be accepted. This is the reason for using several chains which differ only incrementally.

An obvious disadvantage of the algorithm is that m chains are run and only one chain is used for inference. For this reason, MC^3 is ideally suited for implementation on parallel machines, since each chain will in general require the same amount of computation per iteration.

LOCAL Algorithm of Larget and Simon

The LOCAL algorithms offers a computational advantage over previous methods and demonstrates that a Bayesian approach is able to assess uncertainty computationally practical in larger trees. The LOCAL algorithm is an improvement of the GLOBAL algorithm presented in Mau, Newton and Larget in which all branch lengths are changed in every cycle. The LOCAL algorithms modifies the tree by selecting an internal branch of the tree at random. The nodes at the ends of this branch are each connected to two other branches. One of each pair is chosen at random. Imagine taking these three selected edges and stringing them like a clothesline from left to right, where the direction (left/right) is also selected at random. The two endpoints of the first branch selected will have a sub-tree hanging like a piece of clothing strung to the line. The algorithm proceeds by multiplying the three selected branches by a common random amount, akin to stretching or shrinking the clothesline. Finally the leftmost of the two hanging sub-trees is disconnected and reattached to the clothesline at a location selected uniformly at random. This would be the candidate tree.

Suppose we began by selecting the internal branch with length t_8 that separates taxa A and B from the rest. Suppose also that we have (randomly) selected branches with lengths t_1 and t_9 from each side, and that we oriented these branches. Let $m = t_1 + t_8 + t_9$, be the current length of the clothesline. We select the new length to be $m^* = m\exp(\lambda(U_1 - 0.5))$, where U_1 is a uniform random variable on $(0,1)$. Then for the LOCAL algorithm, the acceptance probability can be computed to be:

$$\frac{h(y)}{h(x)} \times \frac{m^{*3}}{m^3}$$

Assessing Convergence

To estimate a branch length t of a 2-taxon tree under JC, in which n_1 sites are unvaried and n_2 are variable, assume exponential prior distribution with rate λ . The density is $p(t) = \lambda e^{-\lambda t}$. The probabilities of the possible site patterns are:

$$1/4\left(1/4+3/4e^{-4/3t}\right)$$

for unvaried sites,

$$1/4\left(1/4-1/4e^{-4/3t}\right)$$

Thus the unnormalized posterior distribution is:

$$h(t) = (1/4)^{n_1+n_2}(1/4+3/4e^{-4/3tn_1})$$

or, alternately,

$$h(t) = \left(1/4-1/4e^{-4/3tn_2}\right)(\lambda e^{-\lambda t})$$

Update branch length by choosing new value uniformly at random from a window of half-width w centered at the current value:

$$t^* = |t + U|$$

where U is uniformly distributed between $-w$ and w. The acceptance probability is:

$$h(t^*)/h(t)$$

Example: $n_1 = 70$, $n_2 = 30$. We will compare results for two values of w, $w = 0.1$ and $w = 0.5$. In each case, we will begin with an initial length of 5 and update the length 2000 times.

Maximum Parsimony and Maximum Likelihood

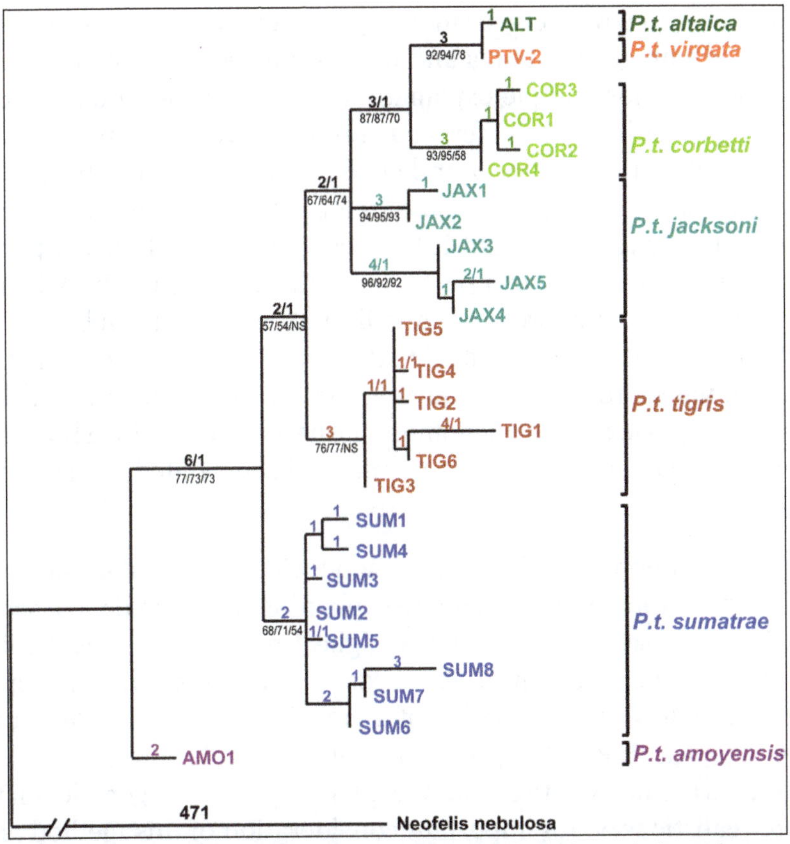

Tiger phylogenetic relationships, bootstrap values shown in branches.

There are many approaches to reconstructing phylogenetic trees, each with advantages and disadvantages, and there is no straightforward answer to "what is the best method?". Maximum parsimony (MP) and maximum likelihood (ML) are traditional methods widely used for the estimation of phylogenies and both use character information directly, as Bayesian methods do.

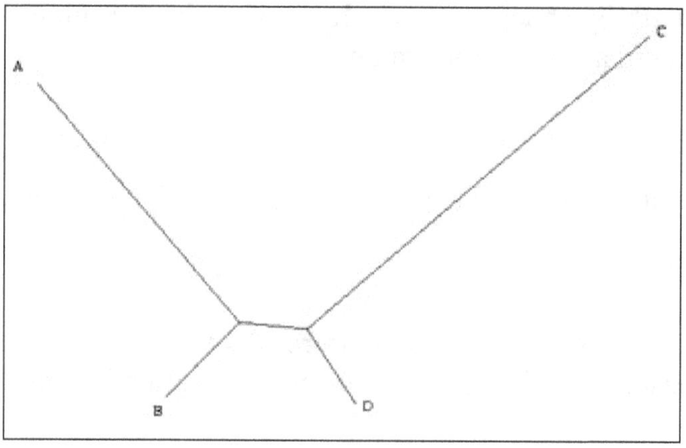

Example of long branch attraction. Longer branches (A & C) appear to be more closely related.

Maximum Parsimony recovers one or more optimal trees based on a matrix of discrete characters for a certain group of taxa and it does not require a model of evolutionary change. MP gives the most simple explanation for a given set of data, reconstructing a phylogenetic tree that includes as few changes across the sequences as possible, this is the one that exhibits the smallest number of evolutionary steps to explain the relationship between taxa. The support of the tree branches is represented by bootstrap percentage. For the same reason that it has been widely used, its simplicity, MP has also received criticism and has been pushed into the background by ML and Bayesian methods. MP presents several problems and limitations. As shown by Felsenstein, MP might be statistically inconsistent, meaning that as more and more data (e.g. sequence length) is accumulated, results can converge on an incorrect tree and lead to long branch attraction, a phylogenetic phenomenon where taxa with long branches (numerous character state changes) tend to appear more closely related in the phylogeny than they really are.

As in maximum parsimony, maximum likelihood will evaluate alternative trees. However it considers the probability of each tree explaining the given data based on a model of evolution. In this case, the tree with the highest probability of explaining the data is chosen over the other ones. In other words, it compares how different trees predict the observed data. The introduction of a model of evolution in ML analyses presents an advantage over MP as the probability of nucleotide substitutions and rates of these substitutions are taken into account, explaining the phylogenetic relationships of taxa in a more realistic way. An important consideration of this method is the branch length, which parsimony ignores, with changes being more likely to happen along long branches than short ones. This approach might eliminate long branch attraction and explain the greater consistency of ML over MP. Although considered by many to be the best approach to inferring phylogenies from a theoretical point of view, ML is computationally intensive and it is almost impossible to explore all trees as there are too many. Bayesian inference also incorporates a model of evolution and the main advantages over MP and ML are that it is computationally more efficient than traditional methods, it quantifies and addresses the source of uncertainty and is able to incorporate complex models of evolution.

Pitfalls and Controversies

- Bootstrap values vs Posterior Probabilities: It has been observed that bootstrap support values, calculated under parsimony or maximum likelihood, tend to be lower than the posterior probabilities obtained by Bayesian inference. This fact leads to a number of questions such as: Do posterior probabilities lead to overconfidence in the results? Are bootstrap values more robust than posterior probabilities?
- Controversy of using prior probabilities: Using prior probabilities for Bayesian analysis has been seen by many as an advantage as it will provide a hypothesis a more realistic view of the real world. However some biologists argue about the subjectivity of Bayesian posterior probabilities after the incorporation of these priors.
- Model choice: The results of the Bayesian analysis of a phylogeny are directly correlated to the model of evolution chosen so it is important to choose a model that fits the observed data, otherwise inferences in the phylogeny will be erroneous. Many scientists have raised questions about the interpretation of Bayesian inference when the model is unknown or incorrect. For example, an oversimplified model might give higher posterior probabilities or simple evolutionary model are associated to less uncertainty than that from bootstrap values.

MrBAYES Software

MrBayes is a free software tool that performs Bayesian inference of phylogeny. Originally written by John P. Huelsenbeck and Frederik Ronquist in 2001. As Bayesian methods increased in popularity MrBayes became one of the software of choice for many molecular phylogeneticists. It is offered for Macintosh, Windows, and UNIX operating systems and it has a command-line interface. The program uses the standard MCMC algorithm as well as the Metropolis coupled MCMC variant. MrBayes reads aligned matrices of sequences (DNA or amino acids) in the standard NEXUS format.

MrBayes uses MCMC to approximate the posterior probabilities of trees. The user can change assumptions of the substitution model, priors and the details of the MC^3 analysis. It also allows the user to remove and add taxa and characters to the analysis. The program uses the most standard model of DNA substitution, the 4x4 also called JC69, which assumes that changes across nucleotides occurs with equal probability. It also implements a number of 20x20 models of amino acid substitution, and codon models of DNA substitution. It offers different methods for relaxing the assumption of equal substitutions rates across nucleotide sites. MrBayes is also able to infer ancestral states accommodating uncertainty to the phylogenetic tree and model parameters.

MrBayes 3 was a completely reorganized and restructured version of the original MrBayes. The main novelty was the ability of the software to accommodate heterogeneity of data sets. This new framework allows the user to mix models and take advantages of the efficiency of Bayesian MCMC analysis when dealing with different type of data (e.g. protein, nucleotide, and morphological). It uses the Metropolis-Coupling MCMC by default.

MrBayes 3.2 new version of MrBayes was released in 2012. The new version allows the users to run multiple analyses in parallel. It also provides faster likelihood calculations and allow these calculations to be delegated to graphics processing unites (GPUs). Version 3.2 provides wider outputs options compatible with FigTree and other tree viewers.

Applications

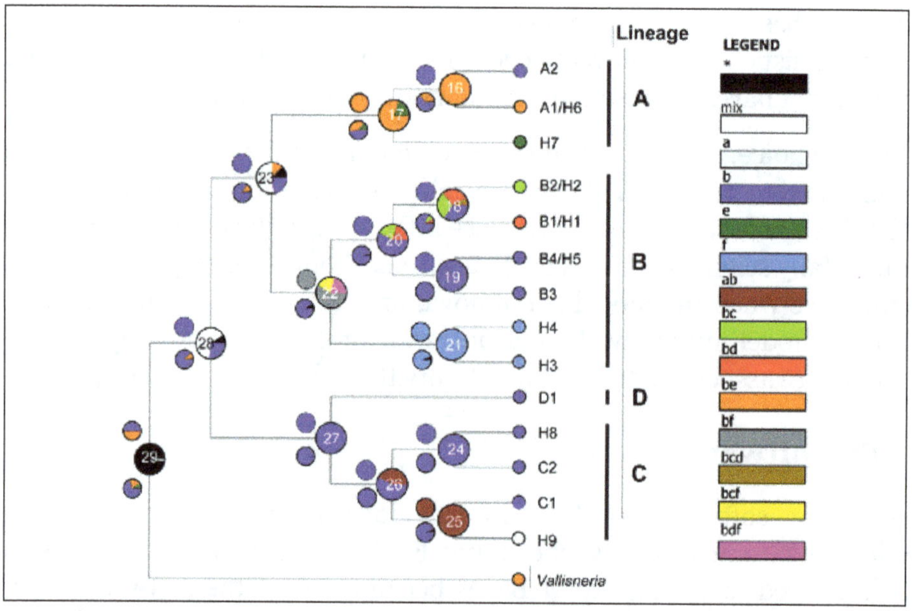

Chronogram obtained from molecular clock analysis using BEAST. Pie chart in each node indicates the possible ancestral distributions inferred from Bayesian Binary MCMC analysis (BBM).

Bayesian Inference has extensively been used by molecular phylogeneticists for a wide number of applications. Some of these include:

- Inference of phylogenies.
- Inference and evaluation of uncertainty of phylogenies.
- Inference of ancestral character state evolution.
- Inference of ancestral areas.
- Molecular dating analysis.

- Model dynamics of species diversification and extinction.
- Elucidate patterns in pathogens dispersal.

Bayesian Information Criterion

In statistics, the Bayesian information criterion (BIC) or Schwarz information criterion (also SIC, SBC, SBIC) is a criterion for model selection among a finite set of models; the model with the lowest BIC is preferred. It is based, in part, on the likelihood function and it is closely related to the Akaike information criterion (AIC).

When fitting models, it is possible to increase the likelihood by adding parameters, but doing so may result in overfitting. Both BIC and AIC attempt to resolve this problem by introducing a penalty term for the number of parameters in the model; the penalty term is larger in BIC than in AIC.

The BIC is formally defined as,

$$\mathrm{BIC} = \ln(n)k - 2\ln(\hat{L}).$$

where,

- $\hat{L}$ = the maximized value of the likelihood function of the model M, i.e. $\hat{L} = p(x \mid \hat{\theta}, M)$, where $\hat{\theta}$ are the parameter values that maximize the likelihood function.
- x = the observed data.
- n = the number of data points in x, the number of observations, or equivalently, the sample size.
- k = the number of parameters estimated by the model. For example, in multiple linear regression, the estimated parameters are the intercept, the q slope parameters, and the constant variance of the errors; thus, $k = q + 2$.

Konishi and Kitagawa derive the BIC to approximate the distribution of the data, integrating out the parameters using Laplace's method, starting with the following:

$$p(x \mid M) = \int p(x \mid \theta, M)\pi(\theta \mid M)d\theta$$

where $\pi(\theta \mid M)$ is the prior for θ under model M.

The log(likelihood), $\ln(p(x \mid \theta, M))$, is then expanded to a second order Taylor series about the MLE, $\hat{\theta}$, assuming it is twice differentiable as follows:

$$\ln(p(x \mid \theta, M)) = \ln(\hat{L}) - 0.5(\theta - \hat{\theta})' n\mathcal{I}(\theta)(\theta - \hat{\theta}) + R(x, \theta),$$

where $\mathcal{I}\theta$ is the average observed information per observation, and prime ($'$) denotes transpose of the vector $(\theta-\hat{\theta})$. To the extent that $R(x,\theta)$ is negligible and $\pi(\theta \mid M)$ is relatively linear near $\hat{\theta}$, we can integrate out θ to get the following:

$$p(x \mid M) \approx \hat{L}(2\pi / n)^{k/2} \mid \mathcal{I}(\hat{\theta}) \mid^{-1/2} \pi(\hat{\theta})$$

As n increases, we can ignore $\mid \mathcal{I}(\hat{\theta}) \mid$ and $\pi(\hat{\theta})$ as they are $O(1)$. Thus,

$$p(x \mid M) = \exp\{\ln \hat{L} - (k/2)\ln(n) + O(1)\} = \exp(-\text{BIC}/2 + O(1)),$$

where BIC is defined as above, and $\hat{L}$ either (a) is the Bayesian posterior mode or (b) uses the MLE and the prior $\pi(\theta \mid M)h$ has nonzero slope at the MLE. Then the posterior

$$p(M \mid x) \propto p(x \mid M)p(M) \approx \exp(-\text{BIC}/2)p(M)$$

Properties

- It is independent of the prior.
- It can measure the efficiency of the parameterized model in terms of predicting the data.
- It penalizes the complexity of the model where complexity refers to the number of parameters in the model.
- It is approximately equal to the minimum description length criterion but with negative sign.
- It can be used to choose the number of clusters according to the intrinsic complexity present in a particular dataset.
- It is closely related to other penalized likelihood criteria such as Deviance information criterion and the Akaike information criterion.

Limitations

The BIC suffers from two main limitations:

- The above approximation is only valid for sample size n much larger than the number k of parameters in the model.
- The BIC cannot handle complex collections of models as in the variable selection (or feature selection) problem in high-dimension.

Gaussian Special Case

Under the assumption that the model errors or disturbances are independent and identically distributed according to a normal distribution and that the boundary condition that the derivative of the log likelihood with respect to the true variance is zero, this becomes (up to an additive constant, which depends only on n and not on the model):

$$\mathrm{BIC} = n\ln(\widehat{\sigma_e^2}) + k\ln(n)$$

where $\widehat{\sigma_e^2}$ is the error variance. The error variance in this case is defined as,

$$\widehat{\sigma_e^2} = \frac{1}{n}\sum_{i=1}^{n}(x_i - \widehat{x_i})^2$$

which is a biased estimator for the true variance.

In terms of the residual sum of squares (RSS) the BIC is,

$$\mathrm{BIC} = n\ln(RSS/n) + k\ln(n)$$

When testing multiple linear models against a saturated model, the BIC can be rewritten in terms of the deviance χ^2 as:

$$\mathrm{BIC} = \chi^2 + k\ln(n)$$

where k is the number of model parameters in the test.

When picking from several models, the one with the lowest BIC is preferred. The BIC is an increasing function of the error variance σ_e^2 and an increasing function of *k*. That is, unexplained variation in the dependent variable and the number of explanatory variables increase the value of BIC. Hence, lower BIC implies either fewer explanatory variables, better fit, or both. The strength of the evidence against the model with the higher BIC value can be summarized as follows:

ΔBIC	Evidence against higher BIC
0 to 2	Not worth more than a bare mention
2 to 6	Positive
6 to 10	Strong
>10	Very strong

The BIC generally penalizes free parameters more strongly than the Akaike information criterion, though it depends on the size of *n* and relative magnitude of *n* and *k*.

It is important to keep in mind that the BIC can be used to compare estimated models only when the numerical values of the dependent variable are identical for all estimates being compared. The models being compared need not be nested, unlike the case when models are being compared using an F-test or a likelihood ratio test.

BIC for High-dimensional Model

For high dimensional model with the number of potential variables $p_n \to \infty$, and the true model size is bounded by a constant, modified BICs has been proposed in Chen and Chen and Gao and Song. For high dimensional model with the number of variables $p_n \to \infty$, and the true model size is unbounded, a high dimensional BIC has been proposed in Gao and Carroll. The high dimensional BIC is of the form:

$$\mathrm{BIC} = 6(1+\gamma)\ln(p_n)k - 2\ln(\hat{L}),$$

where γ can be any number greater than zero.

Gao and Carroll proposed a pseudo-likelihood BIC for which the pseudo log-likelihood is used instead of the true log-likelihood. The high dimensional pseudo-likelihood BIC is of the form:

$$\text{pseudo-BIC} = 6(1+\gamma)\omega\ln(p_n)k^* - 2\ln(\hat{L}),$$

where k^* is an estimated degrees of freedom, and the constant $\omega \geq 1$ is an unknown constant.

To achieve the theoretical model selection consistency for divergent p_n, the two high dimensional BICs above require the multiplicative factor $6(1+\gamma)\omega$. However, in practical use, the high dimensional BIC can take a simpler form:

$$\mathrm{BIC} = c\ln(p_n)k - 2\ln(\hat{L}),$$

where various choices of the multiplicative factor c can be used. In empirical studies, $c = 1$ or $c = 2$ can be used and it is shown to have good empirical performance.

Bayesian Linear Regression

In statistics, Bayesian linear regression is an approach to linear regression in which the statistical analysis is undertaken within the context of Bayesian inference. When the regression model has errors that have a normal distribution, and if a particular form of prior distribution is assumed, explicit results are available for the posterior probability distributions of the model's parameters.

Model Setup

Consider a standard linear regression problem, in which for $i = 1,\ldots,n$ we specify the mean of the conditional distribution of y_i given a $k \times 1$ predictor vector $\mathbf{x}_i$:

$$y_i = \mathbf{x}_i^{\mathrm{T}}\beta + \varepsilon_i,$$

where β is a $k \times 1$ vector, and the ε_i are independent and identically normally distributed random variables:

$$\varepsilon_i \sim N(0,\sigma^2).$$

This corresponds to the following likelihood function:

$$\rho(\mathbf{y} \mid \mathbf{X},\beta,\sigma^2) \propto (\sigma^2)^{-n/2} \exp\left(-\frac{1}{2\sigma^2}(\mathbf{y}-\mathbf{X}\beta)^{\mathrm{T}}(\mathbf{y}-\mathbf{X}\beta)\right).$$

The ordinary least squares solution is used to estimate the coefficient vector using the Moore–Penrose pseudoinverse:

$$\hat{\beta} = (\mathbf{X}^{\mathrm{T}}\mathbf{X})^{-1}\mathbf{X}^{\mathrm{T}}\mathbf{y}$$

where $\mathbf{X}$ is the $n \times k$ design matrix, each row of which is a predictor vector $\mathbf{x}_i^{\mathrm{T}}$; and $\mathbf{y}$ is the column n-vector $[y_1 \cdots y_n]^{\mathrm{T}}$.

This is a frequentist approach, and it assumes that there are enough measurements to say something meaningful about β. In the Bayesian approach, the data are supplemented with additional information in the form of a prior probability distribution. The prior belief about the parameters is combined with the data's likelihood function according to Bayes theorem to yield the posterior belief about the parameters β and σ. The prior can take different functional forms depending on the domain and the information that is available *a priori*.

With Conjugate Priors

Conjugate Prior Distribution

For an arbitrary prior distribution, there may be no analytical solution for the posterior distribution. here, we will consider a so-called conjugate prior for which the posterior distribution can be derived analytically.

A prior $\rho(\beta,\sigma^2)$ is conjugate to this likelihood function if it has the same functional form with respect toâ and σ. Since the log-likelihood is quadratic in β, the log-likelihood is re-written such that the likelihood becomes normal in $(\beta-\hat{\beta})$. Write,

$$(\mathbf{y}-\mathbf{X}\beta)^{\mathrm{T}}(\mathbf{y}-\mathbf{X}\beta) = (\mathbf{y}-\mathbf{X}\hat{\beta})^{\mathrm{T}}(\mathbf{y}-\mathbf{X}\hat{\beta}) + (\beta-\hat{\beta})^{\mathrm{T}}(\mathbf{X}^{\mathrm{T}}\mathbf{X})(\beta-\hat{\beta}).$$

The likelihood is now re-written as,

$$\rho(\boldsymbol{y} \mid \boldsymbol{X},\beta,\sigma^2) \propto (\sigma^2)^{-\frac{v}{2}} \exp(-\frac{vs^2}{2\sigma^2})(\sigma^2)^{-\frac{n-v}{2}} \exp-\left(\frac{1}{2\sigma^2}(\beta-\hat{\beta})^{\mathrm{T}}(\boldsymbol{X}^{\mathrm{T}}\boldsymbol{X})(\beta-\hat{\beta})\right),$$

Where,

$$vs^2 = (\boldsymbol{y} - \boldsymbol{X}\hat{\beta})^{\mathrm{T}}(\boldsymbol{y} - \boldsymbol{X}\hat{\beta}) \quad \text{and} \quad v = n - k,$$

where k is the number of regression coefficients.

This suggests a form for the prior:

$$\rho(\beta,\sigma^2) = \rho(\sigma^2)\rho(\beta \mid \sigma^2),$$

where $\rho(\sigma^2)$ is an inverse-gamma distribution,

$$\rho(\sigma^2) \propto (\sigma^2)^{-\frac{v_0}{2}-1} \exp\left(-\frac{v_0 s_0^2}{2\sigma^2}\right).$$

This is the density of an Inv-Gamma (a_0, b_0) distribution with $a_0 = \frac{v_0}{2} a$ and $b_0 = \frac{1}{2} v_0 s_0^2$ with v_0 and s_0^2 as the prior values of v and s^2, respectively. Equivalently, it can also be described as a scaled inverse chi-squared distribution, Scale-inv-$\chi^2(v_0, s_0^2)$.

Further the conditional prior density $\rho(\beta \mid \sigma^2)$ is a normal distribution,

$$\rho(\beta \mid \sigma^2) \propto (\sigma^2)^{-k/2} \exp\left(-\frac{1}{2\sigma^2}(\beta - \mu_0)^{\mathrm{T}} \Lambda_0 (\beta - \mu_0)\right).$$

In the notation of the normal distribution, the conditional prior distribution is,

$$\mathcal{N}(\mu_0, \sigma^2 \Lambda_0^{-1}).$$

Posterior Distribution

With the prior now specified, the posterior distribution can be expressed as,

$$\rho(\beta,\sigma^2 \mid \boldsymbol{y}, \boldsymbol{X}) \propto \rho(\boldsymbol{y} \mid \boldsymbol{X}, \beta, \sigma^2)\rho(\beta \mid \sigma^2)\rho(\sigma^2)$$

$$\propto (\sigma^2)^{-n/2} \exp\left(-\frac{1}{2\sigma^2}(\boldsymbol{y} - X\beta)^{\mathrm{T}}(\boldsymbol{y} - \boldsymbol{X}\hat{\beta})\right)(\sigma^2)^{-k/2}$$

$$\exp\left(-\frac{1}{2\sigma^2}(\beta - \mu_0)^{\mathrm{T}} \Lambda_0 (\beta - \mu_0)\right)(\sigma^2)^{-(a_0+1)} \exp\left(-\frac{b_0}{\sigma^2}\right)$$

With some re-arrangement, the posterior can be re-written so that the posterior mean μ_n of the parameter vector β can be expressed in terms of the least squares estimator and the prior mean $\hat{\beta}$, with the strength of the prior indicated by the prior precision matrix Λ_0,

$$\mu_n = (\boldsymbol{X}^{\mathrm{T}}\boldsymbol{X} + \Lambda_0)^{-1}(\boldsymbol{X}^{\mathrm{T}}\boldsymbol{X}\hat{\beta} + \Lambda_0 \mu_0).$$

To justify that μ_n is indeed the posterior mean, the quadratic terms in the exponential can be re-arranged as a quadratic form in $\beta - \mu_n$.

$$(\boldsymbol{y}-\boldsymbol{X}\beta)^{\mathrm{T}}(\boldsymbol{y}-\boldsymbol{X}\beta)+(\beta-\mu_0)^{\mathrm{T}}\Lambda_0(\beta-\mu_0)$$
$$=(\beta-\mu_n)^{\mathrm{T}}(\boldsymbol{X}^{\mathrm{T}}\boldsymbol{X}+\Lambda_0)(\hat{\beta}-\mu_n)+\boldsymbol{y}^{\mathrm{T}}\boldsymbol{y}-\mu_n^{\mathrm{T}}(\boldsymbol{X}^{\mathrm{T}}\boldsymbol{X}+\Lambda_0)\mu_n+\mu_0^{\mathrm{T}}\Lambda_0\mu_0.$$

Now the posterior can be expressed as a normal distribution times an inverse-gamma distribution:

$$\rho(\beta,\sigma^2\,|\,\boldsymbol{y},\boldsymbol{X}) \propto (\sigma^2)^{-k/2}\exp\left(-\frac{1}{2\sigma^2}(\beta-\mu_n)^{\mathrm{T}}(\boldsymbol{X}^{\mathrm{T}}\boldsymbol{X}+\Lambda_0)(\beta-\mu_n)\right)$$

$$(\sigma^2)^{-\frac{n+2a_0}{2}-1}\exp\left(-\frac{2b_0+\boldsymbol{y}^{\mathrm{T}}\boldsymbol{y}-\mu_n^{\mathrm{T}}(\boldsymbol{X}^{\mathrm{T}}\boldsymbol{X}+\Lambda_0)\mu_n+\mu_0^{\mathrm{T}}\Lambda_0\mu_0}{2\sigma^2}\right).$$

Therefore, the posterior distribution can be parametrized as follows.

$$\rho(\beta,\sigma^2\,|\,\boldsymbol{y},\boldsymbol{X}) \propto \rho(\beta\,|\,\sigma^2,\boldsymbol{y},\boldsymbol{X})\rho(\sigma^2\,|\,\boldsymbol{y},\boldsymbol{X}),$$

where the two factors correspond to the densities of $\mathcal{N}(\mu_n,\sigma^2\Lambda_n^{-1})$ and Inv-Gamma (a_n,b_n) distributions, with the parameters of these given by,

$$\Lambda_n=(\boldsymbol{X}^{\mathrm{T}}\boldsymbol{X}+\Lambda_0),\quad \mu_n=(\Lambda_n)^{-1}(\boldsymbol{X}^{\mathrm{T}}\boldsymbol{X}\hat{\beta}+\Lambda_0\mu_0),$$

$$a_n=a_0+\frac{n}{2},\qquad b_n=b_0+\frac{1}{2}(\boldsymbol{y}^{\mathrm{T}}\boldsymbol{y}+\mu_0^{\mathrm{T}}\Lambda_0\mu_0-\mu_n^{\mathrm{T}}\Lambda_n\mu_n).$$

This can be interpreted as Bayesian learning where the parameters are updated according to the following equations.

$$\mu_n=(\boldsymbol{X}^{\mathrm{T}}\boldsymbol{X}+\Lambda_0)^{-1}(\Lambda_0\mu_0+\boldsymbol{X}^{\mathrm{T}}\boldsymbol{X}\hat{\beta})=(\boldsymbol{X}^{\mathrm{T}}\boldsymbol{X}+\Lambda_0)^{-1}(\Lambda_0\mu_0+\boldsymbol{X}^{\mathrm{T}}\boldsymbol{y}),$$

$$\Lambda_n=(\boldsymbol{X}^{\mathrm{T}}\boldsymbol{X}+\Lambda_0),$$

$$a_n=a_0+\frac{n}{2},$$

$$b_n=b_0+\frac{1}{2}(\boldsymbol{y}^{\mathrm{T}}\boldsymbol{y}+\mu_0^{\mathrm{T}}\Lambda_0\mu_0-\mu_n^{\mathrm{T}}\Lambda_n\mu_n).$$

Model Evidence

The model evidence $p(\mathbf{y}\,|\,m)$ is the probability of the data given the model m. It is also known as the marginal likelihood, and as the *prior predictive density*. Here, the model is defined by the likelihood function $p(\mathbf{y}\mid\mathbf{X},\beta,\sigma)$ and the prior distribution on the parameters, i.e. $p(\beta,\sigma)$. The model evidence captures in a single number how well such a model explains the observations. The model evidence of the Bayesian linear regression model presented in this topic can be used to compare competing linear models by Bayesian model comparison. These models may differ in the number and values of the predictor variables as well as in their priors on the model parameters. Model complexity is already taken into account by the model evidence, because it marginalizes out the parameters by integrating $p(y,\beta,\sigma\mid X)$ over all possible values of β and σ.

$$p(\mathbf{y}\mid m)=\int p(\mathbf{y}\mid\mathbf{X},\beta,\sigma)p(\beta,\sigma)d\beta d\sigma$$

This integral can be computed analytically and the solution is given in the following equation.

$$p(\boldsymbol{y}\mid m)=\frac{1}{(2\pi)^{n/2}}\sqrt{\frac{\det(\Lambda_0)}{\det(\Lambda_n)}}\cdot\frac{b_0^{a_0}}{b_n^{a_n}}\cdot\frac{\Gamma(a_n)}{\Gamma(a_0)}$$

Here Γ denotes the gamma function. Because we have chosen a conjugate prior, the marginal likelihood can also be easily computed by evaluating the following equality for arbitrary values of â and σ.

$$p(\boldsymbol{y}\mid m)=\frac{p(\beta,\sigma\mid m)p(\boldsymbol{y}\mid\boldsymbol{X},\beta,\sigma,m)}{p(\beta,\sigma\mid\boldsymbol{y},\boldsymbol{X},m)}$$

Note that this equation is nothing but a re-arrangement of Bayes theorem. Inserting the formulas for the prior, the likelihood, and the posterior and simplifying the resulting expression leads to the analytic expression given above.

Other Cases

In general, it may be impossible or impractical to derive the posterior distribution analytically. However, it is possible to approximate the posterior by an approximate Bayesian inference method such as Monte Carlo sampling or variational Bayes.

The special case $\mu_0=0,\Lambda_0=cI$ is called ridge regression.

Bayesian Multivariate Linear Regression

In statistics, Bayesian multivariate linear regression is a Bayesian approach to multivariate linear regression, i.e. linear regression where the predicted outcome is a vector of correlated random variables rather than a single scalar random variable.

Details

Consider a regression problem where the dependent variable to be predicted is not a single real-valued scalar but an m-length vector of correlated real numbers. As in the standard regression setup, there are n observations, where each observation i consists of k-1 explanatory variables, grouped into a vector $\mathbf{x}_i$ of length k (where a dummy variable with a value of 1 has been added to allow for an intercept coefficient). This can be viewed as a set of m related regression problems for each observation i:

$$y_{i,1} = \mathbf{x}_i^{\mathrm{T}} \beta_1 + \epsilon_{i,1}$$

$$\ldots$$

$$y_{i,m} = \mathbf{x}_i^{\mathrm{T}} \beta_m + \epsilon_{i,m}$$

where the set of errors $\{\epsilon_{i,1},\ldots,\epsilon_{i,m}\}$ are all correlated. Equivalently, it can be viewed as a single regression problem where the outcome is a row vector $\mathbf{y}_i^{\mathrm{T}}$ and the regression coefficient vectors are stacked next to each other, as follows:

$$\mathbf{y}_i^{\mathrm{T}} = \mathbf{x}_i^{\mathrm{T}} \mathbf{B} + \epsilon_i^{\mathrm{T}}.$$

The coefficient matrix $\mathbf{B}$ is a $k \times m$ matrix where the coefficient vectors $\beta_1,\ldots,\beta_m$ for each regression problem are stacked horizontally:

$$\mathbf{B} = \begin{bmatrix} \begin{pmatrix} \\ \beta_1 \\ \\ \end{pmatrix} \cdots \begin{pmatrix} \\ \beta_m \\ \\ \end{pmatrix} \end{bmatrix} = \begin{bmatrix} \begin{pmatrix} \beta_{1,1} \\ \vdots \\ \beta_{k,1} \end{pmatrix} \cdots \begin{pmatrix} \beta_{1,m} \\ \vdots \\ \beta_{k,m} \end{pmatrix} \end{bmatrix}.$$

The noise vector ϵ_i for each observation i is jointly normal, so that the outcomes for a given observation are correlated:

$$\epsilon_i \sim N(0, \Sigma_\epsilon).$$

We can write the entire regression problem in matrix form as:

$$\mathbf{Y} = \mathbf{X}\mathbf{B} + \mathbf{E},$$

where $\mathbf{Y}$ and $\mathbf{E}$ are $n \times m$ matrices. The design matrix X is an matrix with the observations $n \times k$ stacked vertically, as in the standard linear regression setup:

$$\mathbf{X} = \begin{bmatrix} \mathbf{x}_1^{\mathrm{T}} \\ \mathbf{x}_2^{\mathrm{T}} \\ \vdots \\ \mathbf{x}_n^{\mathrm{T}} \end{bmatrix} = \begin{bmatrix} x_{1,1} & \cdots & x_{1,k} \\ x_{2,1} & \cdots & x_{2,k} \\ \vdots & \ddots & \vdots \\ x_{n,1} & \cdots & x_{n,k} \end{bmatrix}.$$

The classical, frequentists linear least squares solution is to simply estimate the matrix of regression coefficients $\hat{\mathbf{B}}$ using the Moore-Penrose pseudoinverse:

$$\hat{\mathbf{B}} = (\mathbf{X}^{\mathrm{T}}\mathbf{X})^{-1}\mathbf{X}^{\mathrm{T}}\mathbf{Y}$$

To obtain the Bayesian solution, we need to specify the conditional likelihood and then find the appropriate conjugate prior. As with the univariate case of linear Bayesian regression, we will find that we can specify a natural conditional conjugate prior (which is scale dependent).

Let us write our conditional likelihood as,

$$\rho(\boldsymbol{E} \mid \Sigma_\epsilon) \propto |\Sigma_\epsilon|^{-n/2} \exp(-\frac{1}{2}\mathrm{tr}(\boldsymbol{E}^{\mathrm{T}}\boldsymbol{E}\Sigma_\epsilon^{-1})),$$

writing the error $\mathbf{E}$ in terms of $\mathbf{Y}, \mathbf{X},$ and $\mathbf{B}$ yields,

$$\rho(\boldsymbol{Y} \mid \boldsymbol{X}, \boldsymbol{B}, \Sigma_\epsilon) \propto |\Sigma_\epsilon|^{-n/2} \exp(-\frac{1}{2}\mathrm{tr}((\boldsymbol{Y} - \boldsymbol{XB})^{\mathrm{T}}(\boldsymbol{Y} - \boldsymbol{XB})\Sigma_\epsilon^{-1})),$$

We seek a natural conjugate prior—a joint density $\rho(\mathbf{B}, \Sigma_\epsilon)$ which is of the same functional form as the likelihood. Since the likelihood is quadratic in $\mathbf{B}$, we re-write the likelihood so it is normal in $(\mathbf{B} - \widehat{\mathbf{B}})$ (the deviation from classical sample estimate).

Using the same technique as with Bayesian linear regression, we decompose the exponential term using a matrix-form of the sum-of-squares technique. Here, however, we will also need to use the Matrix Differential Calculus.

First, let us apply sum-of-squares to obtain new expression for the likelihood:

$$\rho(\mathbf{Y} \mid \mathbf{X}, \mathbf{B}, \Sigma_\epsilon) \propto |\Sigma_\epsilon|^{-(n-k)/2} \exp(-\mathrm{tr}(\frac{1}{2}\mathbf{S}^{\mathrm{T}}\mathbf{S}\Sigma_\epsilon^{-1}))|\Sigma_\epsilon|^{-k/2}$$

$$\exp(-\frac{1}{2}\mathrm{tr}((\mathbf{B} - \widehat{\mathbf{B}})^{\mathrm{T}}\mathbf{X}^{\mathrm{T}}\mathbf{X}(\mathbf{B} - \widehat{\mathbf{B}})\Sigma_\epsilon^{-1})),$$

$$\mathbf{S} = \mathbf{Y} - \mathbf{X}\widehat{\mathbf{B}}$$

We would like to develop a conditional form for the priors:

$$\rho(\mathbf{B}, \Sigma_\delta) = \rho(\Sigma_\delta)\rho(\mathbf{B} \mid \Sigma_\delta),$$

where $\rho(\Sigma_\delta)$ is an inverse-Wishart distribution and $\rho(\mathbf{B} \mid \Sigma_\epsilon)$ is some form of normal distribution in the matrix $\mathbf{B}$. This is accomplished using the vectorization transformation, which converts the likelihood from a function of the matrices $\mathbf{B}, \widehat{\mathbf{B}}$ to a function of the vectors $\beta = \mathrm{vec}(\mathbf{B}), \widehat{\beta} = \mathrm{vec}(\widehat{\mathbf{B}})$.

Write,

$$\mathrm{tr}((\mathbf{B}-\widehat{\mathbf{B}})^{\mathrm{T}}\mathbf{X}^{\mathrm{T}}\mathbf{X}(\mathbf{B}-\widehat{\mathbf{B}})\Sigma_{\epsilon}^{-1}) = \mathrm{vec}(\mathbf{B}-\widehat{\mathbf{B}})^{\mathrm{T}}\mathrm{vec}(\mathbf{X}^{\mathrm{T}}\mathbf{X}(\mathbf{B}-\widehat{\mathbf{B}})\Sigma_{\epsilon}^{-1})$$

Let,

$$\mathrm{vec}(\mathbf{X}^{T}\mathbf{X}(\mathbf{B}-\hat{\mathbf{B}})\sum_{\epsilon}^{-1}) = (\sum_{\epsilon}^{-1}\otimes\mathbf{X}^{T}\mathbf{X})\ \mathrm{vec}\ (\mathbf{B}-\hat{\mathbf{B}}),$$

where $\mathbf{A}\otimes\mathbf{B}$ denotes the Kronecker product of matrices A and B, a generalization of the outer product which multiplies an $m\times n$ matrix by a $p\times q$ matrix to generate an $mp\times nq$ matrix, consisting of every combination of products of elements from the two matrices.

Then,

$$\mathrm{vec}(\mathbf{B}-\widehat{\mathbf{B}})^{\mathrm{T}}(\Sigma_{\epsilon}^{-1}\otimes\mathbf{X}^{\mathrm{T}}\mathbf{X})\mathrm{vec}(\mathbf{B}-\widehat{\mathbf{B}})$$

$$=(\beta-\hat{\beta})^{\mathrm{T}}(\Sigma_{\delta}^{-1}\otimes\mathbf{X}^{\mathrm{T}}\mathbf{X})(\beta-\hat{\beta})$$

which will lead to a likelihood which is normal in $(\beta-\hat{\beta})$.

With the likelihood in a more tractable form, we can now find a natural (conditional) conjugate prior.

Conjugate Prior Distribution

The natural conjugate prior using the vectorized variable β is of the form:

$$\rho(\beta,\Sigma_{\epsilon}) = \rho(\Sigma_{\epsilon})\rho(\beta\,|\,\Sigma_{\epsilon}),$$

where,

$$\rho(\Sigma_{\epsilon}) \sim \mathcal{W}^{-1}(V_{\mathbf{o}},\upsilon_{\mathrm{o}})$$

and

$$\rho(\beta\,|\,\Sigma_{\epsilon}) \sim N(\beta_{\mathrm{o}},\Sigma_{\epsilon}\otimes\Lambda_{\mathrm{o}}^{-1}).$$

Posterior Distribution

Using the above prior and likelihood, the posterior distribution can be expressed as:

$$\rho(\beta,\Sigma_{\epsilon}\,|\,\mathbf{Y},\mathbf{X}) \propto |\Sigma_{\epsilon}|^{-(\acute{\imath}_{o}+m+1)/2}\exp(-\frac{1}{2}\mathrm{tr}(\mathbf{V_o}\Sigma_{\epsilon}^{-1}))$$

$$\times|\Sigma_{\epsilon}|^{-k/2}\exp(-\frac{1}{2}\mathrm{tr}((\mathbf{B}-\mathbf{B_o})^{\mathrm{T}}\Lambda_{\mathrm{o}}(\mathbf{B}-\mathbf{B_o})\Sigma_{\epsilon}^{-1}))$$

$$\times |\Sigma_\epsilon|^{-n/2} \exp(-\frac{1}{2}\mathrm{tr}((\mathbf{Y}-\mathbf{XB})^{\mathrm{T}}(\mathbf{Y}-\mathbf{XB})\Sigma_\epsilon^{-1})),$$

where $\mathrm{vec}(\mathbf{B_o}) = \beta_0$. The terms involving $\mathbf{B}$ can be grouped (with $\Lambda_o = \mathbf{U}^{\mathrm{T}}\mathbf{U}$) using:

$$(\mathbf{B}-\mathbf{B_o})^{\mathrm{T}}\Lambda_0(\mathbf{B}-\mathbf{B_o})+(\mathbf{Y}-\mathbf{XB})^{\mathrm{T}}(\mathbf{Y}-\mathbf{XB})$$

$$= \left(\begin{bmatrix}\mathbf{Y}\\ \mathbf{UB_o}\end{bmatrix}-\begin{bmatrix}\mathbf{X}\\ \mathbf{U}\end{bmatrix}\mathbf{B_n}\right)^{\mathrm{T}}\left(\begin{bmatrix}\mathbf{Y}\\ \mathbf{UB_o}\end{bmatrix}-\begin{bmatrix}\mathbf{X}\\ \mathbf{U}\end{bmatrix}\mathbf{B_n}\right)+(\mathbf{B}-\mathbf{B_n})^{\mathrm{T}}(\mathbf{X}^{\mathrm{T}}\mathbf{X}+\Lambda_0)(\mathbf{B}-\mathbf{B_n})$$

$$= (\mathbf{Y}-\mathbf{XB_n})^{\mathrm{T}}(\mathbf{Y}-\mathbf{XB_n})+(\mathbf{B_o}-\mathbf{B_n})^{\mathrm{T}}\Lambda_0(\mathbf{B_o}-\mathbf{B_n})$$

$$+(\mathbf{B}-\mathbf{B_n})^{\mathrm{T}}(\mathbf{X}^{\mathrm{T}}\mathbf{X}+\Lambda_0)(\mathbf{B}-\mathbf{B_n})$$

with,

$$\mathbf{B_n} = (\mathbf{X}^{\mathrm{T}}\mathbf{X}+\Lambda_0)^{-1}(\mathbf{X}^{\mathrm{T}}\mathbf{X}\widehat{\mathbf{B}}+\Lambda_0\mathbf{B_o}) = (\mathbf{X}^{\mathrm{T}}\mathbf{X}+\Lambda_0)^{-1}(\mathbf{X}^{\mathrm{T}}\mathbf{Y}+\Lambda_0\mathbf{B_o}).$$

This now allows us to write the posterior in a more useful form:

$$\rho(\beta,\Sigma_\epsilon \mid \mathbf{Y},\mathbf{X}) \propto |\Sigma_\epsilon|^{-(\mathfrak{i}_0+m+n+1)/2}$$

$$\exp(-\frac{1}{2}\mathrm{tr}((\mathbf{V_o}+(\mathbf{Y}-\mathbf{XB_n})^{\mathrm{T}}(\mathbf{Y}-\mathbf{XB_n})+(\mathbf{B_n}-\mathbf{B_o})^{\mathrm{T}}\Lambda_0(\mathbf{B_n}-\mathbf{B_o}))\Sigma_\epsilon^{-1}))$$

$$\times|\Sigma_\epsilon|^{-k/2}\exp(-\frac{1}{2}\mathrm{tr}((\mathbf{B}-\mathbf{B_n})^{\mathrm{T}}(\mathbf{X}^{T}\mathbf{X}+\Lambda_0)(\mathbf{B}-\mathbf{B_n})\Sigma_\epsilon^{-1}))$$

This takes the form of an inverse-Wishart distribution times a Matrix normal distribution:

$$\rho(\Sigma_\epsilon \mid \boldsymbol{Y},\boldsymbol{X}) \sim \mathcal{W}^{-1}(\boldsymbol{V}_n,\upsilon_n)$$

and

$$\rho(\boldsymbol{B} \mid \boldsymbol{Y},\boldsymbol{X},\Sigma_\delta) \sim \mathcal{MN}_{k,m}(\boldsymbol{B}_n,\Lambda_n^{-1},\Sigma_\delta)$$

The parameters of this posterior are given by:

$$\mathbf{V_n} = \mathbf{V_o}+(\mathbf{Y}-\mathbf{XB_n})^{\mathrm{T}}(\mathbf{Y}-\mathbf{XB_n})+(\mathbf{B_n}-\mathbf{B_o})^{\mathrm{T}}\Lambda_0(\mathbf{B_n}-\mathbf{B_o})$$

$$\upsilon_n = \upsilon_0 + n$$

$$\mathbf{B_n} = (\mathbf{X}^T\mathbf{X} + \Lambda_0)^{-1}(\mathbf{X}^T\mathbf{Y} + \Lambda_0\mathbf{B_0})$$

$$\Lambda_n = \mathbf{X}^T\mathbf{X} + \Lambda_0$$

Bayes Factor

A Bayes factor is the ratio of the likelihood of one particular hypothesis to the likelihood of another. It can be interpreted as a measure of the strength of evidence in favor of one theory among two competing theories.

That's because the Bayes factor gives us a way to evaluate the data in favor of a null hypothesis, and to use external information to do so. It tells us what the weight of the evidence is in favor of a given hypothesis.

When we are comparing two hypotheses, H_1 (the alternate hypothesis) and H_0 (the null hypothesis), the Bayes Factor is often written as B_{10}. It can be defined mathematically as

$$\frac{\textit{liklihood of data given } H_1}{\textit{liklihood of data given } H_0} = \frac{P(D \mid H_1)}{P(D \mid H_0)}$$

The Schwarz criterion is one of the easiest ways to calculate rough approximation of the Bayes Factor.

Interpreting Bayes Factors

A Bayes Factor can be any positive number. One of the most common interpretations is this one—first proposed by Harold Jeffereys (1961) and slightly modified by Lee and Wagenmakers in 2013:

IF B_{10} IS...	THEN YOU HAVE...
> 100	Extreme evidence for H_1
30 – 100	Very strong evidence for H_1
10 – 30	Strong evidence for H_1
3 – 10	Moderate evidence for H_1
1 – 3	Anecdotal evidence for H_1
1	No evidence
1/3 – 1	Anecdotal evidence for H_1
1/3 – 1/10	Moderate evidence for H_1
1/10 – 1/30	Strong evidence for H_1
1/30 – 1/100	Very strong evidence for H_1
< 1/100	Extreme evidence for H_1

References

- Inference-statistics, science: britannica.com, Retrieved 22 July, 2019
- Young, G. A., Smith, R. L. (2005) Essentials of Statistical Inference, CUP. ISBN 0-521-83971-8
- Gelman, Andrew; Carlin, John B.; Stern, Hal S.; Dunson, David B.;Vehtari, Aki; Rubin, Donald B. (2013). Bayesian Data Analysis, Third Edition. Chapman and Hall/CRC. ISBN 978-1-4398-4095-5
- Chen, J.; Chen, Z. (2008). "Extended Bayesian information criteria for model selection with large model spaces". Biometrika. 95 (3): 759–771. CiteSeerX 10.1.1.505.2456. doi:10.1093/biomet/asn034
- Goldstein, Michael; Wooff, David (2007). Bayes Linear Statistics, Theory & Methods. Wiley. ISBN 978-0-470-01562-9
- Bayes-factor-definition: statisticshowto.datasciencecentral.com, Retrieved 09 March, 2019

6

Theorems in Statistics

Central limit theorem, Basu's theorem, Cochran's theorem, Fieller's theorem, Fisher–Tippett–Gnedenko theorem, Hajek–Le Cam convolution theorem, Neyman–Pearson lemma, etc. are some of the theorems that are used in statistics. This chapter discusses these theorems of statistics in detail.

LAW OF LARGE NUMBERS

Law of large numbers, in statistics is the theorem that, as the number of identically distributed, randomly generated variables increases, their sample mean (average) approaches their theoretical mean.

The law of large numbers was first proved by the Swiss mathematician Jakob Bernoulli in 1713. He and his contemporaries were developing a formal probability theory with a view toward analyzing games of chance. Bernoulli envisaged an endless sequence of repetitions of a game of pure chance with only two outcomes, a win or a loss. Labeling the probability of a win p, Bernoulli considered the fraction of times that such a game would be won in a large number of repetitions. It was commonly believed that this fraction should eventually be close to p. This is what Bernoulli proved in a precise manner by showing that, as the number of repetitions increases indefinitely, the probability of this fraction being within any prespecified distance from p approaches 1.

There is also a more general version of the law of large numbers for averages, proved more than a century later by the Russian mathematician Pafnuty Chebyshev.

The law of large numbers is closely related to what is commonly called the law of averages. In coin tossing, the law of large numbers stipulates that the fraction of heads will eventually be close to $^1/_2$. Hence, if the first 10 tosses produce only 3 heads, it seems that some mystical force must somehow increase the probability of a head, producing a return of the fraction of heads to its ultimate limit of $^1/_2$. Yet the law of large numbers requires no such mystical force. Indeed, the fraction of heads can take a very long time to approach $^1/_2$. For example, to obtain a 95 percent probability that the fraction of heads falls between 0.47 and 0.53, the number of tosses must exceed 1,000. In other words, after 1,000 tosses, an initial shortfall of only 3 heads out of 10 tosses is swamped by results of the remaining 990 tosses.

The law of large numbers has a very central role in probability and statistics. It states that if you repeat an experiment independently a large number of times and average the result, what you obtain should be close to the expected value. There are two main versions of the law of large numbers. They are called the weak and strong laws of the large numbers. The difference between them is mostly theoretical.

For i.i.d. random variables $X_1, X_2, \ldots, X_n$ the sample mean, denoted by $\bar{X}$ is defined as,

$$\bar{X} = \frac{X_1 + X_2 + \ldots + X_n}{n}$$

Another common notation for the sample mean is M_n. If the X_i's have CDF $F_X(x)$, we might show the sample mean by $M_n(X)$ to indicate the distribution of the X_i's.

Note that since the X_i's is random variables, the sample mean $\bar{X} = M_n(X)$ is also a random variable. In particular, we have,

$$E[\bar{X}] = \frac{EX_1 + EX_2 + \ldots + EX_n}{n} \quad \text{(by linearity of expectation)}$$

$$= \frac{nEX}{n} \quad \text{(since } EX_i = EX\text{)}$$

$$= EX.$$

Also, the variance of $\bar{X}$ is given by,

$$Var(\bar{X}) = \frac{\mathrm{Var}(X_1 + X_2 + \ldots + X_n)}{n^2} \quad \text{(since Var(aX)} = a^2\text{Var(X))}$$

$$= \frac{\mathrm{Var}(X_1) + \mathrm{Var}(X_2) + \ldots + Var(X_n)}{n^2} \quad \text{(since the } X_i\text{'s are independent)}$$

$$= \frac{nVar(X)}{n^2} \quad \text{(since Var}(X_i) = \text{Var(X))}$$

$$= \frac{Var(X)}{n}.$$

Now let us state and prove the weak law of large numbers (WLLN).

The weak law of large numbers (WLLN).

Let $X_1, X_2, \ldots, X_n$ be i.i.d. random variables with a finite expected value $EX_i = \mu < \infty$. Then, for any $\epsilon > 0$,

$$\lim_{n \to \infty} P(|\bar{X} - \mu| \geq \epsilon) = 0.$$

CENTRAL LIMIT THEOREM

In probability theory, the central limit theorem (CLT) establishes that, in some situations, when independent random variables are added, their properly normalized sum tends toward a normal distribution (informally a "*bell curve*") even if the original variables themselves are not normally distributed. The theorem is a key concept in probability theory because it implies that probabilistic and statistical methods that work for normal distributions can be applicable to many problems involving other types of distributions.

For example, suppose that a sample is obtained containing many observations, each observation being randomly generated in a way that does not depend on the values of the other observations, and that the arithmetic mean of the observed values is computed. If this procedure is performed many times, the central limit theorem says that the distribution of the average will be closely approximated by a normal distribution. A simple example of this is that if one flips a coin many times the probability of getting a given number of heads in a series of flips will approach a normal curve, with mean equal to half the total number of flips in each series; in the limit of an infinite number of flips, it will equal a normal curve.

The central limit theorem has a number of variants. In its common form, the random variables must be identically distributed. In variants, convergence of the mean to the normal distribution also occurs for non-identical distributions or for non-independent observations, given that they comply with certain conditions.

The earliest version of this theorem, that the normal distribution may be used as an approximation to the binomial distribution, is now known as the de Moivre–Laplace theorem.

In more general usage, a central limit theorem is any of a set of weak-convergence theorems in probability theory. They all express the fact that a sum of many independent and identically distributed (i.i.d.) random variables, or alternatively, random variables with specific types of dependence, will tend to be distributed according to one of a small set of *attractor distributions*. When the variance of the i.i.d. variables is finite, the attractor distribution is the normal distribution. In contrast, the sum of a number of i.i.d. random variables with power law tail distributions decreasing as $|x|^{-\alpha-1}$ where $0 < \alpha < 2$ (and therefore having infinite variance) will tend to an alpha-stable distribution with stability parameter (or index of stability) of α as the number of variables grows.

Independent Sequences

Classical CLT

Let $\{X_1, ..., X_n\}$ be a random sample of size n—that is, a sequence of independent and identically distributed (i.i.d.) random variables drawn from a distribution of expected value given by μ and finite variance given by σ^2.

$$S_n := \frac{X_1 + \cdots + X_n}{n}$$

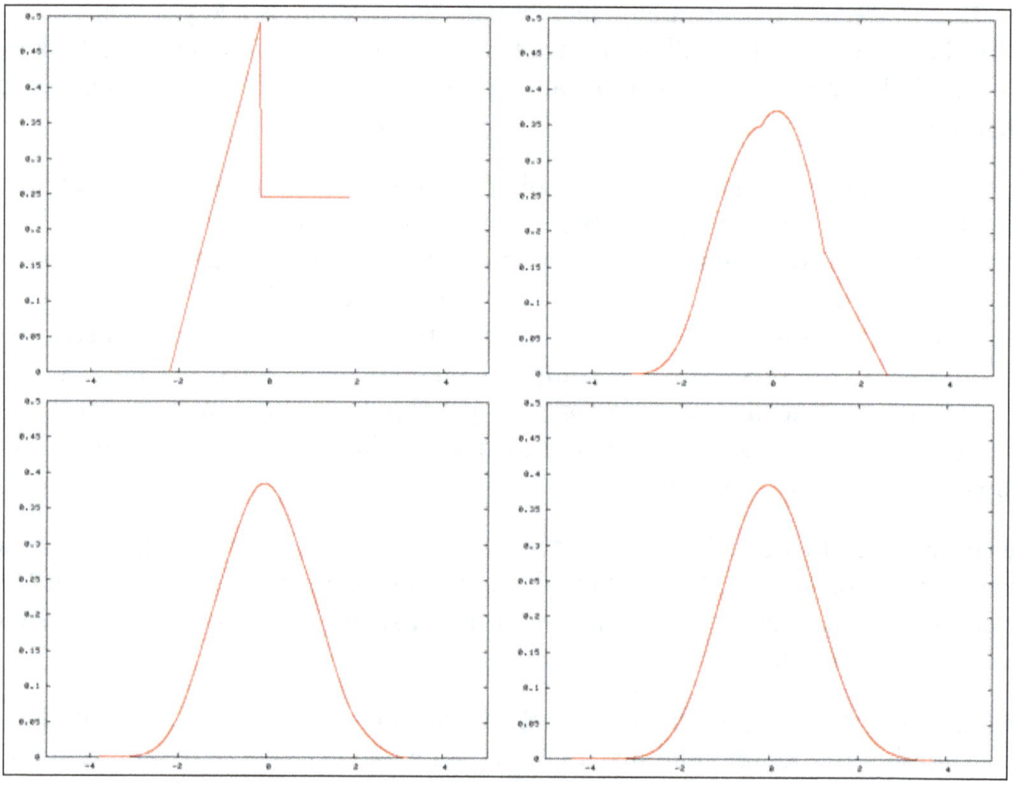

A distribution being "smoothed out" by summation, showing original density of distribution and three subsequent summations.

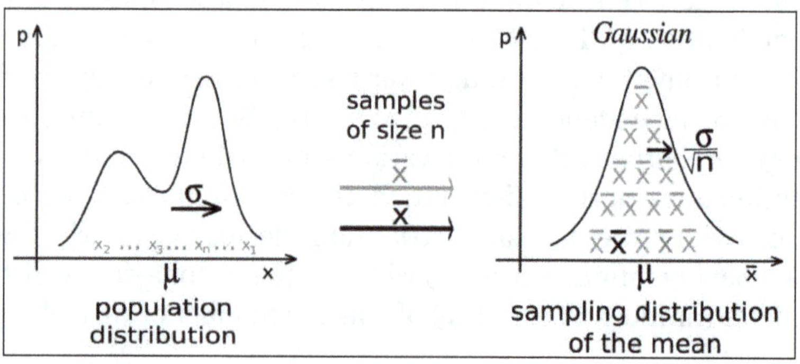

Whatever the form of the population distribution, the sampling distribution tends to a Gaussian, and its dispersion is given by the Central Limit Theorem.

Suppose we are interested in the sample average of these random variables. By the law of large numbers, the sample averages converge in probability and almost surely to the expected value μ as $n \to \infty$. The classical central limit theorem describes the size and the distributional form of the stochastic fluctuations around the deterministic number μ during this convergence. More precisely, it states that as n gets larger, the distribution of the difference between the sample average S_n and its limit μ, when multiplied by the factor $\sqrt{n}$ (*that is* $\sqrt{n}(S_n - \mu)$), approximates the normal distribution with mean 0 and variance σ^2. For large enough n, the distribution of S_n is close to the normal distribution with mean μ and variance σ^2/n. The usefulness of the theorem is that the distribution of $\sqrt{n}(S_n - \mu)$ approaches normality regardless of the shape of the distribution of the individual X_i. Formally, the theorem can be stated as follows:

Lindeberg–Lévy CLT:

Suppose $\{X_1, X_2, \ldots\}$ is a sequence of i.i.d. random variables with $E[X_i] = \mu$ and $Var[X_i] = \sigma^2 < \infty$. Then as n approaches infinity, the random variables $\sqrt{n}(S_n - \mu)$ converge in distribution to a normal $N(0,\sigma^2)$:

$$\sqrt{n}(S_n - \mu) \xrightarrow{d} N(0,\sigma^2).$$

In the case $\sigma > 0$, convergence in distribution means that the cumulative distribution functions of $\sqrt{n}(S_n - \mu)$ converge pointwise to the cdf of the $N(0, \sigma^2)$ distribution: for every real number z,

$$\lim_{n\to\infty} \Pr\left[\sqrt{n}(S_n - \mu) \le z\right] = \lim_{n\to\infty} \Pr\left[\frac{\sqrt{n}(S_n - \mu)}{\sigma} \le \frac{z}{\sigma}\right] = \Phi\left(\frac{z}{\sigma}\right),$$

Where, $\Phi(z)$ is the standard normal cdf evaluated at z. The convergence is uniform in z in the sense that,

$$\lim_{n\to\infty} \sup_{z\in\mathbb{R}} \left| \Pr\left[\sqrt{n}(S_n - \mu) \le z\right] - \Phi\left(\frac{z}{\sigma}\right) \right| = 0,$$

Where, sup denotes the least upper bound (or supremum) of the set.

Lyapunov CLT

The theorem is named after Russian mathematician Aleksandr Lyapunov. In this variant of the central limit theorem the random variables X_i have to be independent, but not necessarily identically distributed. The theorem also requires that random variables $|X_i|$ have moments of some order $(2 + \delta)$, and that the rate of growth of these moments is limited by the Lyapunov condition given below:

Lyapunov CLT: Suppose $\{X_1, X_2, \ldots\}$ is a sequence of independent random variables, each with finite expected value μ_i and variance σ^2_i. Define,

$$s_n^2 = \sum_{i=1}^{n} \sigma_i^2$$

If for some $\delta > 0$, *Lyapunov's condition*,

$$\lim_{n\to\infty} \frac{1}{s_n^{2+\delta}} \sum_{i=1}^{n} \mathrm{E}\left[| X_i - \mu_i |^{2+\delta} \right] = 0$$

is satisfied, then a sum of $\frac{X_i - \mu_i}{s_n}$, converges in distribution to a standard normal random variable, as n goes to infinity:

$$\frac{1}{s_n} \sum_{i=1}^{n} (X_i - \mu_i) \xrightarrow{d} N(0,1).$$

In practice it is usually easiest to check Lyapunov's condition for $\delta = 1$.

If a sequence of random variables satisfies Lyapunov's condition, then it also satisfies Lindeberg's condition. The converse implication, however, does not hold.

Lindeberg CLT

In the same setting and with the same notation as above, the Lyapunov condition can be replaced with the following weaker one.

Suppose that for every $\varepsilon > 0$,

$$\lim_{n\to\infty} \frac{1}{s_n^2} \sum_{i=1}^{n} \mathrm{E}\left[(X_i - \mu_i)^2 \cdot \mathbf{1}_{\{|X_i - \mu_i| > \varepsilon s_n\}} \right] = 0$$

where $\mathbf{1}_{\{|X_i - \mu_i| > \varepsilon s_n\}}$ is the indicator function. Then the distribution of the standardized sums,

$$\frac{1}{s_n} \sum_{i=1}^{n} (X_i - \mu_i)$$

converges towards the standard normal distribution $N(0,1)$.

Multidimensional CLT

Proofs that use characteristic functions can be extended to cases where each individual X_i is a random vector in $\mathbb{R}^k$, with mean vector $\mu = E(X_i)$ and covariance matrix Σ (among the components of the vector), and these random vectors are independent and identically distributed. Summation of these vectors is being done componentwise. The multidimensional central limit theorem states that when scaled, sums converge to a multivariate normal distribution.

Let,

$$X_i = \begin{bmatrix} X_{i(1)} \\ \vdots \\ X_{i(k)} \end{bmatrix}$$

be the k-vector. The bold in X_i means that it is a random vector, not a random (univariate) variable. Then the sum of the random vectors will be,

$$\begin{bmatrix} X_{1(1)} \\ \vdots \\ X_{1(k)} \end{bmatrix} + \begin{bmatrix} X_{2(1)} \\ \vdots \\ X_{2(k)} \end{bmatrix} + \cdots + \begin{bmatrix} X_{n(1)} \\ \vdots \\ X_{n(k)} \end{bmatrix} = \begin{bmatrix} \sum_{i=1}^{n} \left[X_{i(1)} \right] \\ \vdots \\ \sum_{i=1}^{n} \left[X_{i(k)} \right] \end{bmatrix} = \sum_{i=1}^{n} \mathbf{X}_i$$

and the average is,

$$\frac{1}{n} \sum_{i=1}^{n} \mathbf{X}_i = \frac{1}{n} \begin{bmatrix} \sum_{i=1}^{n} X_{i(1)} \\ \vdots \\ \sum_{i=1}^{n} X_{i(k)} \end{bmatrix} = \begin{bmatrix} \bar{X}_{i(1)} \\ \vdots \\ \bar{X}_{i(k)} \end{bmatrix} = \bar{\mathbf{X}}_{\mathbf{n}}$$

and therefore,

$$\frac{1}{\sqrt{n}} \sum_{i=1}^{n} \left[X_i - \mathrm{E}(X_i) \right] = \frac{1}{\sqrt{n}} \sum_{i=1}^{n} (X_i - \mu) = \sqrt{n} \left(\bar{X}_n - \mu \right).$$

The multivariate central limit theorem states that,

$$\sqrt{n} \left(X_n - \mu \right) \xrightarrow{D} N_k(0, \Sigma)$$

where the covariance matrix Σ is equal to,

$$\Sigma = \begin{bmatrix} \mathrm{Var}\left(X_{1(1)}\right) & \mathrm{Cov}\left(X_{1(1)}, X_{1(2)}\right) & \mathrm{Cov}\left(X_{1(1)}, X_{1(3)}\right) & \cdots & \mathrm{Cov}\left(X_{1(1)}, X_{1(k)}\right) \\ \mathrm{Cov}\left(X_{1(2)}, X_{1(1)}\right) & \mathrm{Var}\left(X_{1(2)}\right) & \mathrm{Cov}\left(X_{1(2)}, X_{1(3)}\right) & \cdots & \mathrm{Cov}\left(X_{1(2)}, X_{1(k)}\right) \\ \mathrm{Cov}\left(X_{1(3)}, X_{1(1)}\right) & \mathrm{Cov}\left(X_{1(3)}, X_{1(2)}\right) & \mathrm{Var}\left(X_{1(3)}\right) & \cdots & \mathrm{Cov}\left(X_{1(3)}, X_{1(k)}\right) \\ \vdots & \vdots & \vdots & \ddots & \vdots \\ \mathrm{Cov}\left(X_{1(k)}, X_{1(1)}\right) & \mathrm{Cov}\left(X_{1(k)}, X_{1(2)}\right) & \mathrm{Cov}\left(X_{1(k)}, X_{1(3)}\right) & \cdots & \mathrm{Var}\left(X_{1(k)}\right) \end{bmatrix}.$$

The rate of convergence is given by the following Berry–Esseen type result:

Theorem: Let $X_1, \ldots, X_n$ be independent R^d-valued random vectors, each having mean zero. Write $S = \sum_{i=1}^{n} X_i$ and assume $\Sigma = \mathrm{Cov}[S]$ is invertible. Let $Z \sim N(0, \Sigma)$ be a d-dimensional Gaussian with the same mean and covariance matrix as S. Then for all convex sets $U \subseteq R^d$,

$$|\Pr[S \in U] - \Pr[Z \in U]| \le C d^{1/4} \gamma,$$

where C is a universal constant, $\gamma = \sum_{i=1}^{n} \mathrm{E}[\| \Sigma^{-1/2} X_i \|_2^3]$, , and $\|\cdot\|_2$ denotes the Euclidean norm on R^d.

It is unknown whether the factor $d^{1/4}$ is necessary.

Generalized Theorem

The central limit theorem states that the sum of a number of independent and identically distributed random variables with finite variances will tend to a normal distribution as the number of variables grows. A generalization due to Gnedenko and Kolmogorov states that the sum of a number of random variables with a power-law tail (Paretian tail) distributions decreasing as $|x|^{-\alpha-1}$ where $0 < \alpha < 2$ (and therefore having infinite variance) will tend to a stable distribution $f(x;\alpha,0,c,0)$ as the number of summands grows. If $\alpha > 2$ then the sum converges to a stable distribution with stability parameter equal to 2, i.e. a Gaussian distribution.

Dependent Processes

CLT under Weak Dependence

A useful generalization of a sequence of independent, identically distributed random variables is a mixing random process in discrete time; "mixing" means, roughly, that random variables temporally far apart from one another are nearly independent. Several kinds of mixing are used in ergodic theory and probability theory.

A simplified formulation of the central limit theorem under strong mixing is:

Theorem: Suppose that $X_1, X_2, \ldots$ is stationary and α-mixing with $\alpha_n = O(n^{-5})$ and that $\mathrm{E}(X_n) = 0$ and $\mathrm{E}(X_n^{12}) < \infty$. Denote $S_n = X_1 + \ldots + X_n$, then the limit,

$$\sigma^2 = \lim_n \frac{\mathrm{E}\left(S_n^2\right)}{n}$$

exists, and if $\sigma \neq 0$ then $\frac{S_n}{\sigma\sqrt{n}}$ converges in distribution to $N(0,1)$.

In fact,

$$\sigma^2 = \mathrm{E}\left(X_1^2\right) + 2\sum_{k=1}^{\infty} \mathrm{E}\left(X_1 X_{1+k}\right),$$

where the series converges absolutely.

The assumption $\sigma \neq 0$ cannot be omitted, since the asymptotic normality fails for $X_n = Y_n - Y_{n-1}$ where Y_n are another stationary sequence.

There is a stronger version of the theorem: the assumption $E(X^{12}_n) < \infty$ is replaced with $E(|X_n|^{2+\delta}) < \infty$, and the assumption $\alpha_n = O(n^{-5})$ is replaced with,

$$\sum_n \alpha_n^{\frac{\delta}{2(2+\delta)}} < \infty.$$

Existence of such $\delta > 0$ ensures the conclusion.

Martingale Difference CLT

Theorem: Let a martingale M_n satisfy,

- $\frac{1}{n}\sum_{k=1}^{n} \mathrm{E}\left((M_k - M_{k-1})^2 \mid M_1, \ldots, M_{k-1}\right) \to 1$ in probability as $n \to \infty$,
- For every $\varepsilon > 0$,

  $$\frac{1}{n}\sum_{k=1}^{n} \mathrm{E}\left((M_k - M_{k-1})^2 ; |M_k - M_{k-1}| > \varepsilon\sqrt{n}\right) \to 0$$

  as $n \to \infty$,

  then $M_n/\sqrt{n}$ converges in distribution to $N(0,1)$ as $n \to \infty$.

Caution: The restricted expectation $E(X ; A)$ should not be confused with the conditional expectation $E(X \mid A) = E(X ; A)/ P(A)$.

Proof of Classical CLT

The central limit theorem has a simple proof using characteristic functions. It is similar to the proof of the (weak) law of large numbers.

Assume $\{X_1, \ldots, X_n\}$ are independent and identically distributed random variables, each with mean μ and finite variance σ^2. The sum $X_1 + \ldots + X_n$ has mean $n\mu$ and variance $n\sigma^2$. Consider the random variable,

$$Z_n = \frac{X_1 + \cdots + X_n - n\mu}{\sqrt{n\sigma^2}}$$

$$= \sum_{i=1}^{n} \frac{X_i - \mu}{\sqrt{n\sigma^2}} = \sum_{i=1}^{n} \frac{1}{\sqrt{n}} Y_i,$$

where in the last step we defined the new random variables $Y_i = (X_i - \mu)/\sigma$, each with zero mean and unit variance (var(Y) = 1). The characteristic function of Z_n is given by,

$$\varphi Z_n(t) = \varphi_{\sum_{i=1}^{n}\frac{1}{\sqrt{n}}Y_i}(t) = \varphi Y_1\left(\frac{t}{\sqrt{n}}\right)\varphi Y_2\left(\frac{t}{\sqrt{n}}\right)\cdots\varphi Y_n\left(\frac{t}{\sqrt{n}}\right) = \left[\varphi Y_n\left(\frac{t}{\sqrt{n}}\right)\right]^n,$$

where in the last step we used the fact that all of the Y_i are identically distributed. The characteristic function of Y_1 is, by Taylor's theorem,

$$\varphi Y_1\left(\frac{t}{\sqrt{n}}\right) = 1 - \frac{t^2}{2n} + o\left(\frac{t^2}{n}\right), \quad \left(\frac{t}{\sqrt{n}}\right) \to 0$$

where $o(t^2/n)$ is "little o notation" for some function of t that goes to zero more rapidly than t^2/n. By the limit of the exponential function ($e^x = \lim(1 + x/n)^n$), the characteristic function of Z_n equals,

$$\varphi Z_n(t) = \left(1 - \frac{t^2}{2n} + o\left(\frac{t^2}{n}\right)\right)^n \to e^{-\frac{1}{2}t^2}, \quad n \to \infty.$$

All of the higher order terms vanish in the limit $n \to \infty$. The right hand side equals the characteristic function of a standard normal distribution $N(0,1)$, which implies through Lévy's continuity theorem that the distribution of Z_n will approach $N(0,1)$ as $n \to \infty$. Therefore, the sum $X_1 + \ldots + X_n$ will approach that of the normal distribution $N(n\mu, n\sigma^2)$, and the sample average,

$$S_n = \frac{X_1 + \cdots + X_n}{n}$$

converges to the normal distribution $N(\mu, \sigma^2/n)$, from which the central limit theorem follows.

Convergence to the Limit

The central limit theorem gives only an asymptotic distribution. As an approximation for a finite number of observations, it provides a reasonable approximation only when close to the peak of the normal distribution; it requires a very large number of observations to stretch into the tails.

The convergence in the central limit theorem is uniform because the limiting cumulative distribution function is continuous. If the third central moment $E((X_1 - \mu)^3)$ exists and is finite, then the speed of convergence is at least on the order of $1/\sqrt{n}$. Stein's method can be used not only to prove the central limit theorem, but also to provide bounds on the rates of convergence for selected metrics.

The convergence to the normal distribution is monotonic, in the sense that the entropy of Z_n increases monotonically to that of the normal distribution.

The central limit theorem applies in particular to sums of independent and identically distributed discrete random variables. A sum of discrete random variables is still a discrete random variable, so that we are confronted with a sequence of discrete random variables whose cumulative probability distribution function converges towards a cumulative probability distribution function corresponding to a continuous variable (namely that of the normal distribution). This means that if we build a histogram of the realisations of the sum of n independent identical discrete variables, the curve that joins the centers of the upper faces of the rectangles forming the histogram converges toward a Gaussian curve as n approaches infinity, this relation is known as de Moivre–Laplace theorem.

Relation to the Law of Large Numbers

The law of large numbers as well as the central limit theorem are partial solutions to a general problem: "What is the limiting behaviour of S_n as n approaches infinity?" In mathematical analysis, asymptotic series are one of the most popular tools employed to approach such questions.

Suppose we have an asymptotic expansion of $f(n)$:

$$f(n) = a_1\varphi_1(n) + a_2\varphi_2(n) + O(\varphi_3(n)) \qquad (n \to \infty).$$

Dividing both parts by $\varphi_1(n)$ and taking the limit will produce a_1, the coefficient of the highest-order term in the expansion, which represents the rate at which $f(n)$ changes in its leading term.

$$\lim_{n\to\infty} \frac{f(n)}{\varphi_1(n)} = a_1.$$

Informally, one can say: "$f(n)$ grows approximately as $a_1\varphi_1(n)$". Taking the difference between $f(n)$ and its approximation and then dividing by the next term in the expansion, we arrive at a more refined statement about $f(n)$:

$$\lim_{n\to\infty} \frac{f(n) - a_1\varphi_1(n)}{\varphi_2(n)} = a_2.$$

Here one can say that the difference between the function and its approximation grows approximately as $a_2\varphi_2(n)$. The idea is that dividing the function by appropriate normalizing functions, and looking at the limiting behavior of the result, can tell us much about the limiting behavior of the original function itself.

Informally, something along these lines happens when the sum, S_n, of independent identically distributed random variables, $X_1, \ldots, X_n$, is studied in classical probability theory. If each X_i has finite mean μ, then by the law of large numbers, $S_n/n \to \mu$. If in addition each X_i has finite variance σ^2, then by the central limit theorem,

$$\frac{S_n - n\mu}{\sqrt{n}} \to \xi,$$

where ξ is distributed as $N(0,\sigma^2)$. This provides values of the first two constants in the informal expansion,

$$S_n \approx \mu n + \xi\sqrt{n}.$$

In the case where the X_i do not have finite mean or variance, convergence of the shifted and rescaled sum can also occur with different centering and scaling factors:

$$\frac{S_n - a_n}{b_n} \to \Xi,$$

or informally,

$$S_n \approx a_n + \Xi b_n.$$

Distributions Ξ which can arise in this way are called *stable*. Clearly, the normal distribution is stable, but there are also other stable distributions, such as the Cauchy distribution, for which the mean or variance are not defined. The scaling factor b_n may be proportional to n^c, for any $c \geq 1/2$; it may also be multiplied by a slowly varying function of n.

The law of the iterated logarithm specifies what is happening "in between" the law of large numbers and the central limit theorem. Specifically it says that the normalizing function $\sqrt{n \log \log n}$, intermediate in size between n of the law of large numbers and $\sqrt{n}$ of the central limit theorem, provides a non-trivial limiting behavior.

Alternative Statements of the Theorem

Density Functions

The density of the sum of two or more independent variables is the convolution of their densities (if these densities exist). Thus the central limit theorem can be interpreted as a statement about the properties of density functions under convolution: the convolution of a number of density functions tends to the normal density as the number of density functions increases without bound. These theorems require stronger hypotheses than the forms of the central limit theorem given above. Theorems of this type are often called local limit theorems.

Characteristic Functions

Since the characteristic function of a convolution is the product of the characteristic functions of the densities involved, the central limit theorem has yet another restatement: the product of the characteristic functions of a number of density functions becomes close to the characteristic function of the normal density as the number of density functions increases without bound, under the conditions stated above. Specifically, an appropriate scaling factor needs to be applied to the argument of the characteristic function.

An equivalent statement can be made about Fourier transforms, since the characteristic function is essentially a Fourier transform.

Calculating the Variance

Let S_n be the sum of n random variables. Many central limit theorems provide conditions such that $S_n/\sqrt{Var(S_n)}$ converges in distribution to $N(0,1)$ (the normal distribution with mean 0, variance 1) as $n\to\infty$. In some cases, it is possible to find a constant σ^2 and function $f(n)$ such that $S_n/(\sigma\sqrt{n\cdot f(n)})$ converges in distribution to $N(0,1)$ as $n\to\infty$.

Suppose $X_1, X_2, \ldots$ is a sequence of real-valued and strictly stationary random variables with $\mathbb{E}(X_i)=0$ for all i, $g:[0,1]\to\mathbb{R}$, and $S_n=\sum_{i=1}^{n} g(\frac{i}{n})X_i$. Construct

$$\sigma^2=\mathbb{E}(X_1^2)+2\sum_{i=1}^{\infty}\mathbb{E}(X_1X_{1+i})$$

- If $\sum_{i=1}^{\infty}\mathbb{E}(X_1X_{1+i})$ is absolutely convergent,

  $\left|\int_0^1 g(x)g'(x)dx\right|<\infty$, and $0<\int_0^1 (g(x))^2 dx<\infty$ then $\mathrm{Var}(S_n)/(n\gamma_n)\to\sigma^2$

  as $n\to\infty$

  where $\gamma_n=\frac{1}{n}\sum_{i=1}^{n}(g(\frac{i}{n}))^2$
- If in addition $\sigma>0$ and $S_n/\sqrt{\mathrm{Var}(S_n)}$ converges in distribution to $\mathcal{N}(0,1)$ as $n\to\infty$ then $S_n/(\sigma\sqrt{n\gamma_n})$ also converges in distribution to $\mathcal{N}(0,1)$ as $n\to\infty$.

Extensions

Products of Positive Random Variables

The logarithm of a product is simply the sum of the logarithms of the factors. Therefore, when the logarithm of a product of random variables that take only positive values approaches a normal distribution, the product itself approaches a log-normal distribution. Many physical quantities (especially mass or length, which are a matter of scale and cannot be negative) are the products of different random factors, so they follow a log-normal distribution. This multiplicative version of the central limit theorem is sometimes called Gibrat's law.

Whereas the central limit theorem for sums of random variables requires the condition of finite variance, the corresponding theorem for products requires the corresponding condition that the density function be square-integrable.

Beyond the Classical Framework

Asymptotic normality, that is, convergence to the normal distribution after appropriate shift and rescaling, is a phenomenon much more general than the classical framework treated above, namely, sums of independent random variables (or vectors). New frameworks are revealed from time to time; no single unifying framework is available for now.

Convex Body

Theorem: There exists a sequence $\varepsilon_n \downarrow 0$ for which the following holds. Let $n \geq 1$, and let random variables $X_1, \ldots, X_n$ have a log-concave joint density f such that $f(x_1, \ldots, x_n) = f(|x_1|, \ldots, |x_n|)$ for all $x_1, \ldots, x_n$, and $E(X^2_k) = 1$ for all $k = 1, \ldots, n$. Then the distribution of,

$$\frac{X_1 + \cdots + X_n}{\sqrt{n}}$$

is ε_n-close to $N(0,1)$ in the total variation distance.

These two ε_n-close distributions have densities (in fact, log-concave densities), thus, the total variance distance between them is the integral of the absolute value of the difference between the densities. Convergence in total variation is stronger than weak convergence.

An important example of a log-concave density is a function constant inside a given convex body and vanishing outside; it corresponds to the uniform distribution on the convex body, which explains the term "central limit theorem for convex bodies".

Another example: $f(x_1, \ldots, x_n) = \text{const} \cdot \exp(-(|x_1|^\alpha + \ldots + |x_n|^\alpha)^\beta)$ where $\alpha > 1$ and $\alpha\beta > 1$. If $\beta = 1$ then $f(x_1, \ldots, x_n)$ factorizes into $\text{const} \cdot \exp(-|x_1|^\alpha) \ldots \exp(-|x_n|^\alpha)$, which means $X_1, \ldots, X_n$ are independent. In general, however, they are dependent.

The condition $f(x_1, \ldots, x_n) = f(|x_1|, \ldots, |x_n|)$ ensures that $X_1, \ldots, X_n$ are of zero mean and uncorrelated; still, they need not be independent, nor even pairwise independent. By the way, pairwise independence cannot replace independence in the classical central limit theorem.

Here is a Berry–Esseen type result.

Theorem: Let $X_1, \ldots, X_n$ satisfy the assumptions of the previous theorem, then,

$$\left|\mathbb{P}\left(a \le \frac{X_1 + \cdots + X_n}{\sqrt{n}} \le b\right) - \frac{1}{\sqrt{2\pi}}\int_a^b e^{-\frac{1}{2}t^2}\,dt\right| \le \frac{C}{n}$$

for all $a < b$; here C is a universal (absolute) constant. Moreover, for every $c_1, \ldots, c_n \in \mathbb{R}$ such that $c^2_1 + \ldots + c^2_n = 1$,

$$\left|\mathbb{P}(a \le c_1X_1 + \cdots + c_nX_n \le b) - \frac{1}{\sqrt{2\pi}}\int_a^b e^{-\frac{1}{2}t^2}\,dt\right| \le C\left(c_1^4 + \ldots + c_n^4\right).$$

The distribution of $(X_1 + \ldots + X_n)/\sqrt{n}$ need not be approximately normal (in fact, it can be uniform). However, the distribution of $c_1X_1 + \ldots + c_nX_n$ is close to $N(0,1)$ (in the total variation distance) for most vectors $(c_1, \ldots, c_n)$ according to the uniform distribution on the sphere $c^2_1 + \ldots + c^2_n = 1$.

Lacunary Trigonometric Series

Theorem: (Salem–Zygmund): Let U be a random variable distributed uniformly on $(0,2\pi)$, and $X_k = r_k \cos(n_kU + a_k)$, where,

- n_k satisfy the lacunarity condition: there exists $q > 1$ such that $n_{k+1} \ge qn_k$ for all k,
- r_k are such that,

  $$r_1^2 + r_2^2 + \cdots = \infty$$

  and

  $$\frac{r_k^2}{r_1^2 + \cdots + r_k^2} \to 0,$$

- $0 \le a_k < 2\pi$.

Then,

$$\frac{X_1 + \cdots + X_k}{\sqrt{r_1^2 + \cdots + r_k^2}}$$

converges in distribution to $N(0, 1/2)$.

Gaussian Polytopes

Theorem: Let $A_1, \ldots, A_n$ be independent random points on the plane $\mathbb{R}^2$ each having the two-dimensional standard normal distribution. Let K_n be the convex hull of these points, and X_n the area of K_n. Then,

$$\frac{X_n - \mathrm{E}(X_n)}{\sqrt{\mathrm{Var}(X_n)}}$$

converges in distribution to $N(0,1)$ as n tends to infinity.

The same also holds in all dimensions greater than 2.

The polytope K_n is called a Gaussian random polytope.

A similar result holds for the number of vertices (of the Gaussian polytope), the number of edges, and in fact, faces of all dimensions.

Linear Functions of Orthogonal Matrices

A linear function of a matrix M is a linear combination of its elements (with given coefficients), $\mathrm{M} \mapsto \mathrm{tr}(\mathrm{AM})$ where A is the matrix of the coefficients;

A random orthogonal matrix is said to be distributed uniformly, if its distribution is the normalized Haar measure on the orthogonal group $\mathrm{O}(n, \mathbb{R})$.

Theorem: Let M be a random orthogonal $n \times n$ matrix distributed uniformly, and A a fixed $n \times n$ matrix such that $\mathrm{tr}(\mathrm{AA}^*) = n$, and let $X = \mathrm{tr}(\mathrm{AM})$. Then the distribution of X is close to $N(0,1)$ in the total variation metric up to $2\sqrt{3}/n - 1$.

Subsequences

Theorem: Let random variables $X_1, X_2, \ldots \in L_2(\Omega)$ be such that $X_n \to 0$ weakly in $L_2(\Omega)$ and $X_n \to 1$ weakly in $L_1(\Omega)$. Then there exist integers $n_1 < n_2 < \ldots$ such that,

$$\frac{X_{n_1} + \cdots + X_{n_k}}{\sqrt{k}}$$

converges in distribution to $N(0,1)$ as k tends to infinity.

Random Walk on a Crystal Lattice

The central limit theorem may be established for the simple random walk on a crystal lattice (an infinite-fold abelian covering graph over a finite graph), and is used for design of crystal structures.

Applications and Examples

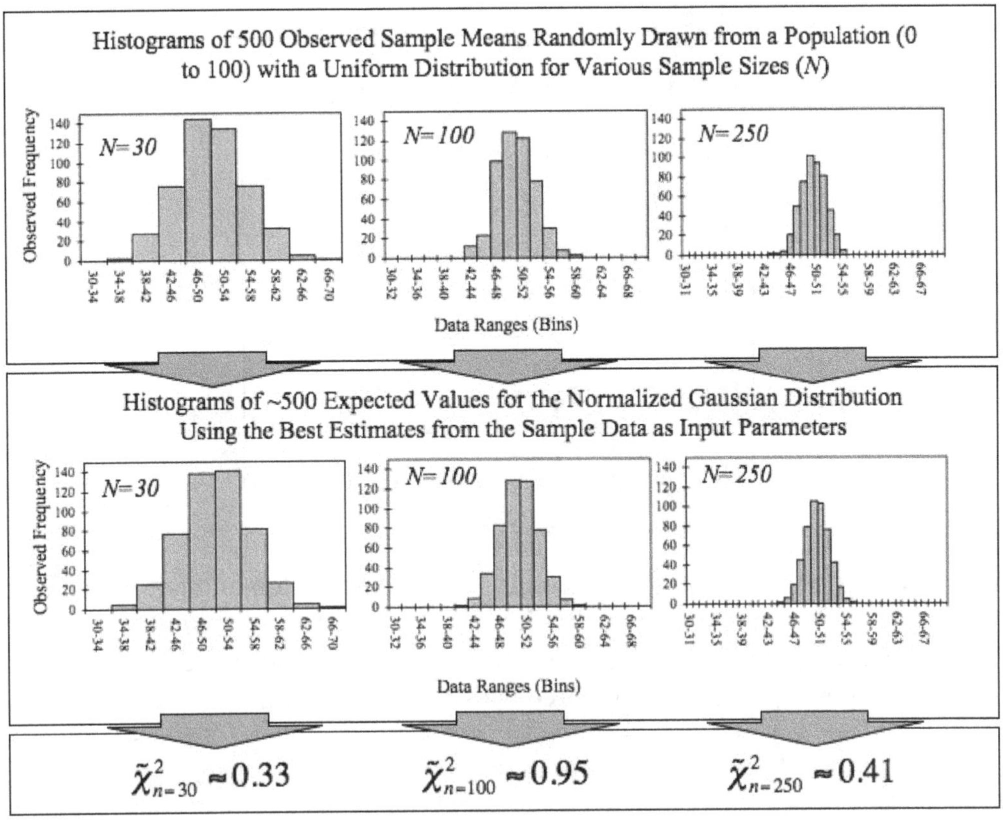

This figure demonstrates the central limit theorem. The sample means are generated using a random number generator, which draws numbers between 0 and 100 from a uniform probability distribution. It illustrates that increasing sample sizes result in the 500 measured sample means being more closely distributed about the population mean (50 in this case). It also compares the observed distributions with the distributions that would be expected for a normalized Gaussian distribution, and shows the chi-squared values that quantify the goodness of the fit (the fit is good if the reduced chi-squared value is less than or approximately equal to one). The input into the normalized Gaussian function is the mean of sample means (~50) and the mean sample standard deviation divided by the square root of the sample size ($\sim 28.87/\sqrt{n}$), which is called the standard deviation of the mean (since it refers to the spread of sample means).

A simple example of the central limit theorem is rolling many identical, unbiased dice. The distribution of the sum (or average) of the rolled numbers will be well approximated by a normal distribution. Since real-world quantities are often the balanced sum of many unobserved random events, the central limit theorem also provides a partial explanation for the prevalence of the normal probability distribution. It also justifies the approximation of large-sample statistics to the normal distribution in controlled experiments.

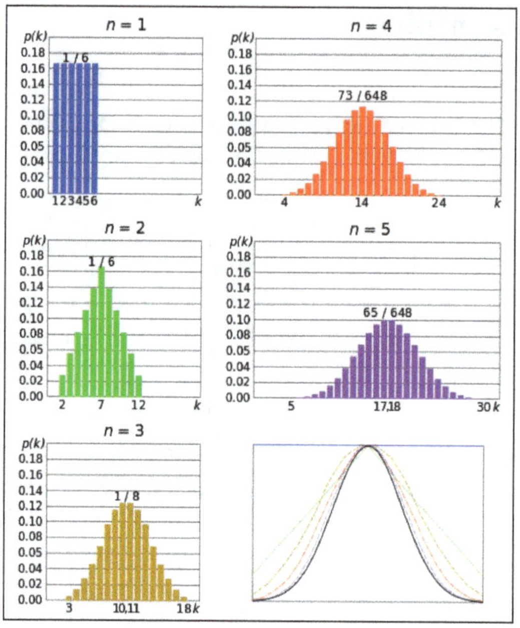

Comparison of probability density functions, **$p(k)$ for the sum of n fair 6-sided dice to show their convergence to a normal distribution with increasing n, in accordance to the central limit theorem. In the bottom-right graph, smoothed profiles of the previous graphs are rescaled, superimposed and compared with a normal distribution (black curve).

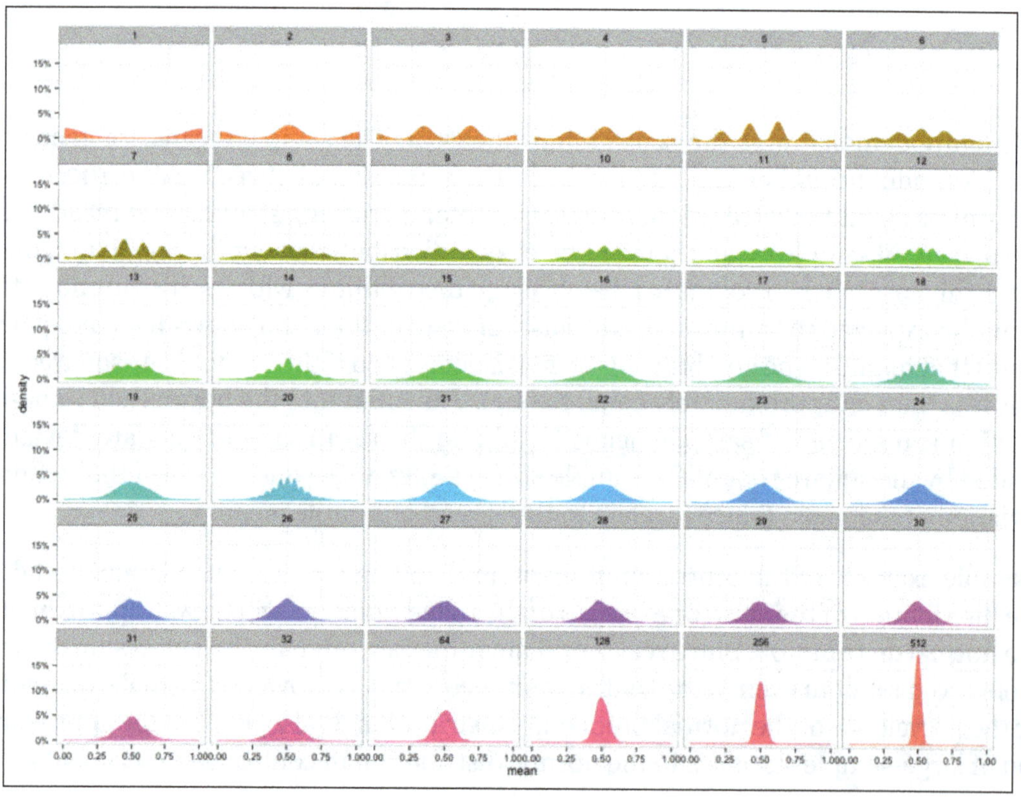

Another simulation using the binomial distribution. Random 0s and 1s were generated, and then their means calculated for sample sizes ranging from 1 to 512. Note that as the sample size increases the tails become thinner and the distribution becomes more concentrated around the mean.

Real Applications

Published literature contains a number of useful and interesting examples and applications relating to the central limit theorem. One source states the following examples:

- The probability distribution for total distance covered in a random walk (biased or unbiased) will tend toward a normal distribution.
- Flipping many coins will result in a normal distribution for the total number of heads (or equivalently total number of tails).

From another viewpoint, the central limit theorem explains the common appearance of the "bell curve" in density estimates applied to real world data. In cases like electronic noise, examination grades, and so on, we can often regard a single measured value as the weighted average of many small effects.

Using generalisations of the central limit theorem, we can then see that this would often (though not always) produce a final distribution that is approximately normal.

In general, the more a measurement is like the sum of independent variables with equal influence on the result, the more normality it exhibits. This justifies the common use of this distribution to stand in for the effects of unobserved variables in models like the linear model.

Regression

Regression analysis and in particular ordinary least squares specifies that a dependent variable depends according to some function upon one or more independent variables, with an additive error term. Various types of statistical inference on the regression assume that the error term is normally distributed.

This assumption can be justified by assuming that the error term is actually the sum of many independent error terms; even if the individual error terms are not normally distributed, by the central limit theorem their sum can be well approximated by a normal distribution.

Other Illustrations

Given its importance to statistics, a number of papers and computer packages are available that demonstrate the convergence involved in the central limit theorem.

BASU'S THEOREM

In statistics, Basu's theorem states that any boundedly complete minimal sufficient statistic is independent of any ancillary statistic. This is a 1955 result of Debabrata Basu.

It is often used in statistics as a tool to prove independence of two statistics, by first demonstrating one is complete sufficient and the other is ancillary, then appealing to the theorem. An example of this is to show that the sample mean and sample variance of a normal distribution are independent statistics, This property (independence of sample mean and sample variance) characterizes normal distributions.

Statement:

Let $(P_\theta;\theta \in \Theta)$ be a family of distributions on a measurable space $(X,\mathcal{A})$ and T, A measurable maps from $(X,\mathcal{A})$ to some measurable space $(Y,\mathcal{B})$. (Such maps are called a statistic.) If T is a boundedly complete sufficient statistic for θ, and A is ancillary to θ, then T is independent of A.

Proof:

Let P_θ^T and P_θ^A be the marginal distributions of T and A respectively.

Denote by $A^{-1}(B)$ the preimage of a set B under the map A. For any measurable set $B \in \mathcal{B}$ we have,

$$P_\theta^A(B) = P_\theta(A^{-1}(B)) = \int_Y P_\theta(A^{-1}(B) \mid T = t)\, P_\theta^T(dt).$$

The distribution P_θ^A does not depend on θ because A is ancillary. Likewise, $P_\theta(\cdot \mid T = t)$ does not depend on θ because T is sufficient. Therefore,

$$\int_Y Y[P(A^{-1}(B) \mid T = t) - P^A(B)]\, P_\theta^T(dt) = 0.$$

Note the integrand (the function inside the integral) is a function of t and not θ. Therefore, since T is boundedly complete the function,

$$g(t) = P(A^{-1}(B) \mid T = t) - P^A(B)$$

is zero for P_θ^T almost all values of t and thus,

$$P(A^{-1}(B) \mid T = t) = P^A(B)$$

for almost all t. Therefore, A is independent of T.

Independence of Sample Mean and Sample Variance of a Normal Distribution (Known Variance)

Let $X_1, X_2, ..., X_n$ be independent, identically distributed normal random variables with mean μ and variance σ^2.

Then with respect to the parameter μ, one can show that,

$$\hat{\mu} = \frac{\Sigma X_i}{n},$$

the sample mean, is a complete sufficient statistic – it is all the information one can derive to estimate μ, and no more – and

$$\hat{\sigma}^2 = \frac{\sum\left(X_i - \bar{X}\right)^2}{n-1},$$

the sample variance, is an ancillary statistic – its distribution does not depend on μ.

Therefore, from Basu's theorem it follows that these statistics are independent.

This independence result can also be proven by Cochran's theorem.

Further, this property (that the sample mean and sample variance of the normal distribution are independent) characterizes the normal distribution – no other distribution has this property.

COCHRAN'S THEOREM

In statistics, Cochran's theorem, devised by William G. Cochran, is a theorem used to justify results relating to the probability distributions of statistics that are used in the analysis of variance.

Statement:

Suppose $U_1, ..., U_N$ are i.i.d. standard normally distributed random variables, and there exist matrices $B^{(1)}, B^{(2)}, \ldots, B^{(k)}$, with $\sum_{i=1}^{k} B^{(i)} = I_N$.

Further suppose that $r_1 + \cdots + r_k = N$,, where r_i is the rank of $B^{(i)}$. If we write,

$$Q_i = \sum_{j=1}^{N} \sum_{\ell=1}^{N} U_j B_{j,\ell}^{(i)} U_\ell$$

so that the Q_i are quadratic forms, then Cochran's theorem states that the Q_i are independent, and each Q_i has a chi-squared distribution with r_i degrees of freedom.

Less formally, it is the number of linear combinations included in the sum of squares defining Q_i, provided that these linear combinations are linearly independent.

Proof:

We first show that the matrices $B^{(i)}$ can be simultaneously diagonalized and that their non-zero eigenvalues are all equal to +1. We then use the vector basis that diagonalize them to simplify their characteristic function and show their independence and distribution.

Each of the matrices $B^{(i)}$ has rank r_i and thus r_i non-zero eigenvalues. For each i, the sum $C^{(i)} \equiv \sum_{j \neq i} B^{(j)}$ has at most rank $\sum_{j \neq i} r_j = N - r_i$. Since

$B^{(i)} + C^{(i)} = I_{N \times N}$, it follows that $C^{(i)}$ has exactly rank $N - r_i$.

Therefore $B^{(i)}$ and $C^{(i)}$ can be simultaneously diagonalized. This can be shown by first diagonalizing $B^{(i)}$. In this basis, it is of the form:

$$\begin{bmatrix} \lambda_1 & 0 & 0 & \cdots & \cdots & & 0 \\ 0 & \lambda_2 & 0 & \cdots & \cdots & & 0 \\ 0 & 0 & \ddots & & & & \vdots \\ \vdots & \vdots & & \lambda_{r_i} & & & \\ \vdots & \vdots & & & 0 & & \\ 0 & \vdots & & & & \ddots & \\ 0 & 0 & \cdots & & & & 0 \end{bmatrix}.$$

Thus the lower $(N - r_i)$ rows are zero. Since $C^{(i)} = I - B^{(i)}$, it follows that these rows in $C^{(i)}$ in this basis contain a right block which is a $(N - r_i) \times (N - r_i)$ unit matrix, with zeros in the rest of these rows. But since $C^{(i)}$ has rank $N - r_i$, it must be zero elsewhere. Thus it is diagonal in this basis as well. It follows that all the non-zero eigenvalues of both $B^{(i)}$ and $C^{(i)}$ are +1. Moreover, the above analysis can be repeated in the diagonal basis for $C^{(1)} = B^{(2)} + \sum_{j>2} B^{(j)}$. In this basis $C^{(1)}$ is the identity of an $(N - r_1) \times (N - r_1)$ vector space, so it follows that both $B^{(2)}$ and $\sum_{j>2} B^{(j)}$ are simultaneously diagonalizable in this vector space (and hence also together with $B^{(1)}$). By iteration it follows that all B-s are simultaneously diagonalizable.

Thus there exists an orthogonal matrix S such that for all i, $S^{\mathrm{T}} B^{(i)} S \equiv B^{(i)\prime}$ is diagonal, where any entry $B_{x,y}^{(i)}$ ' with indices $x=y$, $\sum_{j=1}^{i-1} r_j < x = y \le \sum_{j=1}^{i} r_j$, is equal to 1, while any entry with other indices is equal to 0.

Let U_i' denote some specific linear combination of all U_i after transformation by S. Note that $\sum_{i=1}^{N}(U_i')^2 = \sum_{i=1}^{N} U_i^2$ due to the length preservation of the orthogonal matrix S, that the Jacobian of a linear transformation is the matrix associated with the linear transformation itself, and that the determinant of an orthogonal matrix has modulus 1.

The characteristic function of Q_i is:

$$\varphi_i(t) = (2\pi)^{-N/2}\int du_1 \int du_2 \cdots \int du_N e^{itQ_i} \cdot e^{-u_1^2/2} \cdot e^{-u_2^2/2} \cdots e^{-u_N^2/2}$$

$$= (2\pi)^{-N/2}\left(\prod_{j=1}^{N}\int du_j\right) e^{itQ_i} \cdot e^{-\sum_{j=1}^{N} u_j^2/2}$$

$$= (2\pi)^{-N/2}\left(\prod_{j=1}^{N}\int du_j'\right) e^{it\cdot\sum_{m=r_1+\cdots+r_{i-1}+1}^{r_1+\cdots+r_i}(u_m')^2} \cdot e^{-\sum_{j=1}^{N} u_j'^2/2}.$$

$$= (2\pi)^{-N/2}\left(\int e^{u^2(it-\frac{1}{2})}du\right)^{r_i}\left(\int e^{-\frac{u^2}{2}}du\right)^{N-r_i}$$

$$= (1-2it)^{-r_i/2}$$

This is the Fourier transform of the chi-squared distribution with r_i degrees of freedom. Therefore this is the distribution of Q_i.

Moreover, the characteristic function of the joint distribution of all the Q_is is:

$$\varphi(t_1,t_2,\ldots,t_k) = (2\pi)^{-N/2}\left(\prod_{j=1}^{N}\int dU_j\right) e^{i\sum_{i=1}^{k} t_i\cdot Q_i} \cdot e^{-\sum_{j=1}^{N} U_j^2/2}$$

$$= (2\pi)^{-N/2}\left(\prod_{j=1}^{N}\int dU_j'\right) e^{i\cdot\sum_{i=1}^{k} t_i \sum_{k=r_1+\cdots+r_{i-1}+1}^{r_1+\cdots+r_i}(U_k')^2} \cdot e^{-\sum_{j=1}^{N} U_j'^2/2}$$

$$= (2\pi)^{-N/2}\prod_{i=1}^{k}\left(\int e^{u^2(it_i-\frac{1}{2})}du\right)^{r_i}$$

$$= \prod_{i=1}^{k}(1-2it_i)^{-r_i/2} = \prod_{i=1}^{k}\varphi_i(t_i)$$

From this it follows that all the Q_is are independent.

Sample Mean and Sample Variance

If $X_1, ..., X_n$ are independent normally distributed random variables with mean μ and standard deviation σ then,

$$U_i = \frac{X_i - \mu}{\sigma}$$

is standard normal for each i. It is possible to write,

$$\sum_{i=1}^{n} U_i^2 = \sum_{i=1}^{n}\left(\frac{X_i - \bar{X}}{\sigma}\right)^2 + n\left(\frac{\bar{X} - \mu}{\sigma}\right)^2$$

(here $\bar{X}$ is the sample mean). To see this identity, multiply throughout by σ^2 and note that,

$$\sum (X_i - \mu)^2 = \sum (X_i - \bar{X} + \bar{X} - \mu)^2$$

and expand to give,

$$\sum (X_i - \mu)^2 = \sum (X_i - \bar{X})^2 + \sum (\bar{X} - \mu)^2 + 2\sum (X_i - \bar{X})(\bar{X} - \mu).$$

The third term is zero because it is equal to a constant times,

$$(\bar{X} - X_i) = 0,$$

and the second term has just n identical terms added together. Thus,

$$\sum (X_i - \mu)^2 = \sum (X_i - \bar{X})^2 + n(\bar{X} - \mu)^2,$$

and hence,

$$\sum\left(\frac{X_i - \mu}{\sigma}\right)^2 = \sum\left(\frac{X_i - \bar{X}}{\sigma}\right)^2 + n\left(\frac{\bar{X} - \mu}{\sigma}\right)^2 = Q_1 + Q_2.$$

Now the rank of $B^{(2)}$ is just 1 (it is the square of just one linear combination of the standard normal variables). The rank of $B^{(1)}$ can be shown to be $n - 1$, and thus the conditions for Cochran's theorem are met.

Cochran's theorem then states that Q_1 and Q_2 are independent, with chi-squared distributions with $n - 1$ and 1 degree of freedom respectively. This shows that the sample mean and sample variance are independent. This can also be shown by Basu's theorem, and in fact this property *characterizes* the normal distribution – for no other distribution are the sample mean and sample variance independent.

Distributions

The result for the distributions is written symbolically as,

$$\sum\left(X_i-\bar{X}\right)^2 \sim \sigma^2\chi^2_{n-1}.$$

$$n(\bar{X}-\mu)^2 \sim \sigma^2\chi^2_1,$$

Both these random variables are proportional to the true but unknown variance σ^2. Thus their ratio does not depend on σ^2 and, because they are statistically independent. The distribution of their ratio is given by,

$$\frac{n\left(\bar{X}-\mu\right)^2}{\frac{1}{n-1}\sum\left(X_i-\bar{X}\right)^2} \sim \frac{\chi^2_1}{\frac{1}{n-1}\chi^2_{n-1}} \sim F_{1,n-1}$$

where $F_{1,n-1}$ is the F-distribution with 1 and $n-1$ degrees of freedom. The final step here is effectively the definition of a random variable having the F-distribution.

Estimation of Variance

To estimate the variance σ^2, one estimator that is sometimes used is the maximum likelihood estimator of the variance of a normal distribution,

$$\hat{\sigma}^2 = \frac{1}{n}\sum\left(X_i-\bar{X}\right)^2.$$

Cochran's theorem shows that,

$$\frac{n\hat{\sigma}^2}{\sigma^2} \sim \chi^2_{n-1}$$

and the properties of the chi-squared distribution show that,

$$E\left(\frac{n\hat{\sigma}^2}{\sigma^2}\right) = E\left(\chi^2_{n-1}\right)$$

$$\frac{n}{\sigma^2}E\left(\hat{\sigma}^2\right) = (n-1)$$

$$E\left(\hat{\sigma}^2\right) = \frac{\sigma^2(n-1)}{n}$$

Alternative Formulation

The following version is often seen when considering linear regression. Suppose that $Y \sim N_n(0, \sigma^2 I_n)$ is a standard multivariate normal random vector (here I_n denotes the n-by-n identity matrix), and if $A_1, \ldots, A_k$ are all n-by-n symmetric matrices with $\sum_{i=1}^{k} A_i = I_n$. Then, on defining $r_i = Rank(A_i)$, any one of the following conditions implies the other two:

- $\sum_{i=1}^{k} r_i = n,$
- $Y^T A_i Y \sim \sigma^2 \chi^2_{r_i}$ (thus the A_i are positive semidefinite),
- $Y^T A_i Y$ is independent of $Y^T A_j Y$ for $i \neq j$.

FIELLER'S THEOREM

In statistics, Fieller's theorem allows the calculation of a confidence interval for the ratio of two means.

Approximate Confidence Interval

Variables a and b may be measured in different units, so there is no way to directly combine the standard errors as they may also be in different units. The most complete discussion of this is given by Fieller.

Fieller showed that if a and b are (possibly correlated) means of two samples with expectations μ_a and μ_b, and variances $v_{11}\sigma^2$ and $v_{22}\sigma^2$ and covariance $v_{12}\sigma^2$, and if v_{11}, v_{12}, v_{22} are all known, then a (1 – α) confidence interval (m_L, m_U) for μ_a / μ_b is given by,

$$(m_L, m_U) = \frac{1}{(1-g)}\left[\frac{a}{b} - \frac{g v_{12}}{v_{22}} \mp \frac{t_{r,\alpha} s}{b}\sqrt{v_{11} - 2\frac{a}{b}v_{12} + \frac{a^2}{b^2}v_{22} - g\left(v_{11} - \frac{v_{12}^2}{v_{22}}\right)}\right]$$

where,

$$g = \frac{t_{r,\alpha}^2 s^2 v_{22}}{b^2}.$$

Here s^2 is an unbiased estimator of σ^2 based on r degrees of freedom, and $t_{r,\alpha}$ is the α -level deviate from the Student's t-distribution based on r degrees of freedom.

Three features of this formula are important in this context:

- The expression inside the square root has to be positive, or else the resulting interval will be imaginary.
- When g is very close to 1, the confidence interval is infinite.
- When g is greater than 1, the overall divisor outside the square brackets is negative and the confidence interval is exclusive.

Other Methods

One problem is that, when g is not small, the confidence interval can blow up when using Fieller's theorem. Andy Grieve has provided a Bayesian solution where the CIs are still sensible, albeit wide. Bootstrapping provides another alternative that does not require the assumption of normality.

FISHER–TIPPETT–GNEDENKO THEOREM

Ronald Fisher.

In statistics, the Fisher–Tippett–Gnedenko theorem (also the Fisher–Tippett theorem or the extreme value theorem) is a general result in extreme value theory regarding asymptotic distribution of extreme order statistics. The maximum of a sample of iid random variables after proper renormalization can only converge in distribution to one of 3 possible distributions, the Gumbel distribution, the Fréchet distribution, or the Weibull distribution. Credit for the extreme value theorem (or convergence to types theorem) is given to Gnedenko, previous versions were stated by Ronald Fisher and Leonard Henry Caleb Tippettin 1928 and Fréchet in 1927.

The role of the extremal types theorem for maxima is similar to that of central limit theorem for averages, except that the central limit theorem applies to the average of a sample from any distribution with finite variance, while the Fisher–Tippet–Gnedenko theorem only states that if the distribution of a normalized maximum converges, then the limit has to be one of a particular class of distributions. It does not state that the distribution of the normalized maximum does converge.

Statement:

Let $X_1, X_2 \ldots, X_n \ldots$ be a sequence of independent and identically-distributed random variables, and $M_n = \max\{X_1, \ldots, X_n\}$.. If a sequence of pairs of real numbers (a_n, b_n) exists such that each $a_n > 0$ and $\lim_{n\to\infty} P\left(\frac{M_n - b_n}{a_n} \le x\right) = F(x)$, where F is a non-degenerate distribution function, then the limit distribution F belongs to either the Gumbel, the Fréchet or the Weibull family. These can be grouped into the generalized extreme value distribution.

Conditions of Convergence

If G is the distribution function of X, then M_n can be rescaled to converge in distribution to

- A Fréchet if and only if $G(x) < 1$ for all real x and $\frac{1-G(tx)}{1-G(t)} \xrightarrow[t\to+\infty]{} x^{-\theta}, \quad x > 0.$ In this case, possible sequences are,

  bn = 0 and $a_n = G^{-1}\left(1 - \frac{1}{n}\right).$

- A Weibull if and only if $\omega = \sup\{G < 1\} < +\infty,$

- $\frac{1-G(\omega+tx)}{1-G(\omega-t)} \xrightarrow[t\to 0^+]{} (-x)^{\theta}, \quad x < 0$ In this case possible sequences are,

  bn = ω and $a_n = \omega - G^{-1}\left(1 - \frac{1}{n}\right).$

Convergence conditions for the Gumbel distribution are more involved.

GAUSS–MARKOV THEOREM

The Gauss-Markov theorem states that if your linear regression model satisfies the first six classical assumptions, then ordinary least squares (OLS) regression produces unbiased estimates that have the smallest variance of all possible linear estimators.

The Gauss-Markov Theorem: OLS is BLUE

The Gauss-Markov theorem famously states that OLS is BLUE. BLUE is an acronym for the following:

Best Linear Unbiased Estimator

The definition of "best" refers to the minimum variance or the narrowest sampling distribution. More specifically, when your model satisfies the assumptions, OLS coefficient estimates follow the tightest possible sampling distribution of unbiased estimates compared to other linear estimation methods.

What does OLS Estimate?

Regression analysis is like any other inferential methodology. Our goal is to draw a random sample from a population and use it to estimate the properties of that population. In regression analysis, the coefficients in the equation are estimates of the actual population parameters.

The notation for the model of a population is the following:

$$Y = \beta_0 + \beta_1 X_1 + \beta_2 X_2 + \cdots + \beta_k X_k + \epsilon$$

The hats over the betas indicate that these are parameter estimates while e represents the residuals, which are estimates of the random error.

Typically, statisticians consider estimates to be useful when they are unbiased (correct on average) and precise (minimum variance). To apply these concepts to parameter estimates and the Gauss-Markov theorem, we'll need to understand the sampling distribution of the parameter estimates.

Sampling Distributions of the Parameter Estimates

Imagine that we repeat the same study many times. We collect random samples of the same size, from the same population, and fit the same OLS regression model repeatedly. Each random sample produces different estimates for the parameters in the regression equation. After this process, we can graph the distribution of estimates for each parameter. Statisticians refer to this type of distribution as a sampling distribution, which is a type of probability distribution.

Keep in mind that each curve represents the sampling distribution of the estimates for a single parameter. The graphs below tell us which values of parameter estimates are more and less common. They also indicate how far estimates are likely to fall from the correct value.

HÁJEK–LE CAM CONVOLUTION THEOREM

In statistics, the Hájek–Le Cam convolution theorem states that any regular estimator in a parametric model is asymptotically equivalent to a sum of two independent random variables, one of which is normalwith asymptotic variance equal to the inverse of Fisher information, and the other having arbitrary distribution.

The obvious corollary from this theorem is that the "best" among regular estimators are those with the second component identically equal to zero. Such estimators are called efficient and are known to always exist for regular parametric models.

The theorem is named after Jaroslav Hájek and Lucien Le Cam.

Statement:

Let $\wp = \{P_\theta \mid \theta \in \Theta \subset \mathrm{R}^k\}$ be a regular parametric model, and $q(\theta)\colon \Theta \to \mathrm{R}^m$ be a parameter in this model (typically a parameter is just one of the components of vector θ). Assume that function q is differentiable on Θ, with the $m \times k$ matrix of derivatives denoted as $\dot{q}_\theta$. Define,

$$I_{q(\theta)}^{-1} = \dot{q}(\theta) I^{-1}(\theta) \dot{q}(\theta)' \text{ — the information bound for q,}$$

$$\psi_{q(\theta)} = \dot{q}(\theta) I^{-1}(\theta) \dot{\ell}(\theta) \text{ — the efficient influence function for q,}$$

where $I(\theta)$ is the Fisher information matrix for model $\wp$, $\dot{\ell}(\theta)$ is the score function, and$'$ denotes matrix transpose.

Theorem: Suppose T_n is a uniformly (locally) regular estimator of the parameter q. Then,

- There exist independent random m-vectors $Z_\theta \sim \mathcal{N}(0, I_{q(\theta)}^{-1})$ and Δ_θ such that,

  $$\sqrt{n}(T_n - q(\theta)) \xrightarrow{d} Z_\theta + \Delta_\theta,$$

  where ^{d} denotes convergence in distribution. More specifically,

  $$\begin{pmatrix} \sqrt{n}(T_n - q(\theta)) - \frac{1}{\sqrt{n}} \sum_{i=1}^{n} \psi_{q(\theta)}(x_i) \\ \frac{1}{\sqrt{n}} \sum_{i=1}^{n} \psi_{q(\theta)}(x_i) \end{pmatrix} \xrightarrow{d} \begin{pmatrix} \Delta_\theta \\ Z_\theta \end{pmatrix}.$$

- If the map $\theta \to \dot{q}_\theta$ is continuous, then the convergence in (A) holds uniformly on compact subsets of Θ. Moreover, in that case $\Delta_\theta = 0$ for all θ if and only if T_n is uniformly (locally) asymptotically linear with influence function $\psi_{q(\theta)}$.

LEHMANN–SCHEFFÉ THEOREM

In statistics, the Lehmann–Scheffé theorem is a prominent statement, tying together the ideas of completeness, sufficiency, uniqueness, and best unbiased estimation. The theorem states that any estimator which is unbiased for a given unknown quantity and that depends on the data only through a complete, sufficient statistic is the unique best unbiased estimator of that quantity. The Lehmann–Scheffé theorem is named after Erich Leo Lehmann and Henry Scheffé, given their two early papers.

If T is a complete sufficient statistic for θ and $E(g(T)) = \tau(\theta)$ then $g(T)$ is the uniformly minimum-variance unbiased estimator (UMVUE) of $\tau(\theta)$.

Statement:

Let $\vec{X} = X_1, X_2, \ldots, X_n$ be a random sample from a distribution that has p.d.f (or p.m.f in the discrete case) $f(x:\theta)$ where $\theta \in \Omega$ is a parameter in the parameter space. Suppose $Y = u(\vec{X})$ is a sufficient statistic for θ, and let $\{f_Y(y:\theta) : \theta \in \Omega\}$ be a complete family. If $\varphi : \mathrm{E}[\varphi(Y)] = \theta$ then $\varphi(Y)$ is the unique MVUE of θ.

Proof:

By the Rao–Blackwell theorem, if Z is an unbiased estimator of θ then $\varphi(Y) := \mathrm{E}[Z \mid Y]$ defines an unbiased estimator of θ with the property that its variance is not greater than that of Z.

Now we show that this function is unique. Suppose W is another candidate MVUE estimator of θ. Then again $\psi(Y) := \mathrm{E}[W \mid Y]$ defines an unbiased estimator of θ with the property that its variance is not greater than that of W. Then,

$$\mathrm{E}[\varphi(Y) - \psi(Y)] = 0, \theta \in \Omega.$$

Since $\{f_Y(y:\theta) : \theta \in \Omega\}$ is a complete family,

$$\mathrm{E}[\varphi(Y) - \psi(Y)] = 0 \Rightarrow \varphi(y) - \psi(y) = 0, \theta \in \Omega$$

and therefore the function φ is the unique function of Y with variance not greater than that of any other unbiased estimator. We conclude that $\varphi(Y)$ is the MVUE.

Example for when using a Non-complete Minimal Sufficient Statistic

An example of an improvable Rao–Blackwell improvement, when using a minimal sufficient statistic that is not complete, was provided by Galili and Meilijson in 2016. Let $X_1, \ldots, X_n$ be a random sample from a scale-uniform distribution $X \sim U((1-k)\theta, (1+k)\theta)$, with unknown mean $\mathrm{E}[X] = \theta$ and known design parameter

$k \in (0,1)$. In the search for "best" possible unbiased estimators for θ, it is natural to consider X_1 as an initial (crude) unbiased estimator for θ and then try to improve it. Since X_1 is not a function of $T = \left(X_{(1)}, X_{(n)}\right)$, the minimal sufficient statistic for θ (where $X_{(1)} = \min_i X_i$ and $X_{(n)} = \max_i X_i \beta$), it may be improved using the Rao–Blackwell theorem as follows:

$$\hat{\theta}_{RB} = \mathrm{E}_\theta[X_1 \mid X_{(1)}, X_{(n)}] = \frac{X_{(1)} + X_{(n)}}{2}.$$

However, the following unbiased estimator can be shown to have lower variance:

$$\hat{\theta}_{LV} = \frac{1}{k^2 \dfrac{n-1}{n+1} + 1} \cdot \frac{(1-k)X_{(1)} + (1+k)X_{(n)}}{2}.$$

And in fact, it could be even further improved when using the following estimator:

$$\hat{\theta}_{\text{BAYES}} = \frac{n+1}{n}\left[1 - \frac{\dfrac{X_{(1)}(1+k)}{X_{(n)}(1-k)} - 1}{\left(\dfrac{X_{(1)}(1+k)}{X_{(n)}(1-k)}\right)^{n+1} - 1}\right]\frac{X_{(n)}}{1+k}$$

NEYMAN–PEARSON LEMMA

Neyman Pearson Lemma is used for testing a statistical hypothesis to test whether the performed test is the most powerful test about the population parameter with the consideration of the supposed probability distribution.

It allows seeing whether the rejection region which has been selected is the best one or not. It helps to assess the statistical power of the hypothesis test. The statistical power of the hypothesis test states that the null hypothesis has been correctly rejected in favor of the alternative hypothesis.

A test with the highest power of all the tests for the same level of significance is called the most-powerful test. Suppose if the results of the observations are used to test the null hypothesis as against the simple alternative hypothesis, the error arises from the rejection of null hypothesis being verified, as per a statistical test formulated to test a null hypothesis against the alternative hypothesis if the null hypothesis is actually true.

The most powerful tests are constructed by Neyman-Pearson lemma. As per this, the most powerful test is the likelihood-ratio. A test proposed for testing the simple null hypothesis against a simple alternative hypothesis which offers the least probability of error among all the tests is the most powerful test. As the statistical test power is obtained by subtracting the probability of a type II error by one, the most powerful test is formulated in terms of probabilities of errors of type I and type II errors.

Neyman-Pearson lemma test is defined as below:

Assumetherandomsamples $Y_1, Y_2, \ldots Y_n$ withparameterarefromaprobabilitydistribution. Then if the critical region is D of size 'a' and a constant k such that:

$$\frac{L(\theta_o)}{L(\theta_\infty)} \leq k \qquad \text{inside the critical religion D,}$$

$$\frac{L(\theta_o)}{L(\theta_\infty)} \geq k \qquad \text{outside the critical religion D.}$$

Here, D is the most powerful critical region and hence the most powerful test.

References

- Law-of-large-numbers, science: britannica.com, Retrieved 15 June, 2019
- Bárány, Imre; Vu, Van (2007). "Central limit theorems for Gaussian polytopes". Annals of Probability. Institute of Mathematical Statistics. 35 (4): 1593–1621. arXiv:math/0610192. doi:10.1214/009117906000000791
- Casella, George (2001). Statistical Inference. Duxbury Press. p. 369. ISBN 978-0-534-24312-8
- O'Hagan A, Stevens JW, Montmartin J (2000). "Inference for the cost-effectiveness acceptability curve and cost-effectiveness ratio". Pharmacoeconomics. 17 (4): 339–49. doi:10.2165/00019053-200017040-00004. PMID 10947489
- Gauss-markov-theorem-ols-blue, regression: statisticsbyjim.com, Retrieved 06 February, 2019
- Neyman-pearson-lemma-31: chegg.com, Retrieved 28 March, 2019

7

Applications

Statistics has applications in the fields of actuarial science, business analytics, forensics, finance, engineering, operations research, signal processing, psychology, machine learning, etc. This chapter has been carefully written to provide an easy understanding of the diverse applications of statistics.

BUSINESS STATISTICS

Business statistics takes the data analysis tools from elementary statistics and applies them to business. For example, estimating the probability of a defect coming off a factory line, or seeing where sales are headed in the future. Many of the tools used in business statistics are built on ones you've probably already come across in basic math: mean, mode and median, bar graphs and the bell curve, and basic probability. Hypothesis testing (where you test out an idea) and regression analysis (fitting data to an equation) builds on this foundation.

Describing Populations and Samples

The process of describing populations and samples is called Descriptive Statistics. A population includes everyone in the area of interest. For example, every person in the United States, every dog owner in Florida, or every computer user in the world. A sample is a small piece of the whole (i.e. 1000 people in the United States, 250 Floridian dog owners, 2500 worldwide computer users). There are three main ways to describe populations and samples: central tendency, dispersion and association.

Measures of Central Tendency

In this area, you find where the bulk of the data lies. It includes finding the mean, mode and median. The formulas for finding the population mean and sample mean have slightly different symbols:

Sample mean formula: $\bar{x} = (\Sigma x_i)/n$.

Population mean: $\mu = (\Sigma * X)/N$.

They are solved in the same way: add the items together and then divide by the number of items in the set.

Measures of Dispersion

How much is your data set spread out around the mean? Is there a big difference between your highest and lowest values? These are questions that can be answered by finding the interquartile range, variance and standard deviation. The interquartile range is especially useful if you are more interested in where the bulk of your data lies and less interested in extreme values. For example, a business geared towards 20-somethings might want to plot the age range for customers who walk in the door. This makes sure they are marketing to the right age group; ideally, most customers should be in their 20s.

Measures of Association

Measures of association tell you about trends in data. For example, you could make a plot showing current manufacturing costs. This might show a high or low connection ("correlation") between different factors and final cost. The factors could include employee time off, the price of oil, or location of the plant. Covariance is how two variables change together. If the price of tomatoes goes up, it directly affects the price of ketchup. The price of corn (to make high fructose corn syrup) also affects the price of ketchup, but in a smaller way.

Probabilities and Random Variables

Probability is the foundation of business statistics. Several formulas are used, including the basic formula:

P(A) = number of outcomes that give A/number of possible outcomes = r/n.

A simple example: A box of factory rejects contains 5 balls that are too small, 3 balls that are too big and 2 under-inflated balls. If a ball is chosen at random from the box, what is the probability that it is: (a) too small; (b) too small or under-inflated; (c) not under-inflated?

Answers:

a. P(A) = r/n = 5/10 = 0.5;

b. P(A) = r/n = 5+2/10 = 0.7;

c. P(A) = r/n = 8/10 = 0.8.

There are dozens of ways to figure out probabilities. It largely depends on what you want to know. For example, something happening or not happening, choosing people or items.

In algebra, and "x" or "y" can represent a number, like 3,14 or 22.5. In statistics, a random variable must be linked to a random event or experiment. Let's say you wanted to know how many faulty televisions are produced on a certain line. Your random variable, X could be the number of faulty televisions produced in 24 hours. A business decision model takes data and applies logic to arrive at a business decision. In the case of the faulty televisions, data could be measured each time a change is made to the production line. This data could be used to see what changes lead to improve quality, and which do not.

Probability Distributions

Probability distributions can be discrete or continuous.

Discrete Distributions

Examples of discrete distributions include the Binomial Distribution.

In a binomial experiment, there are only two outcomes (like yes/no or success/failure). The formula is:

$$P(X)=\frac{n!}{(n-X)!X!}\cdot(p)^{X}\cdot(q)^{n-X}$$

An employment test has 10 multiple choice questions with five choices for each question. If someone guesses randomly (i.e. without reading the questions), what is the probability they get exactly 6 questions right?

Plugging the values into the formula, you get:

$$P(X) = 10! / (10 - 6)!\ 6! * (.2)^{6} * (.8)^{10-6} = 0.005505024$$

The "p" here is 0.2 because you have a 1 out of 5, or 20% chance of getting a question right if you guess. "q" is just 1-p (100% – 20% = 80%). You might actually see "1-P" in some versions of this formula.

Similar Distributions to the binomial:

- The Negative Binomial distribution is similar to the binomial, but it does not have a fixed number of trials.
- The Geometric Distribution represents the number of failures before you get a success in a series of Bernoulli trials.
- Poisson Distribution: Used to predict the probability of certain events from happening when you know how often the event has occurred. For example, if a store that rents books has an average rental of 200 books every Saturday night. Using this data, you can predict how many more books will sell on the following Saturday nights.

Continuous probability distributions can take on an infinite number of different values. Examples include:

- Normal distribution: Commonly known as the "bell curve", it's used to model lots of situations, like exam scores, IQ, and heights.
- Student's T distribution: Similar to the normal, but used for small samples (under 30 items).
- Chi-square distribution: Has many used in stats, like comparing categorical variables.
- F Distribution: Used for a specific type of test called Analysis of Variance.

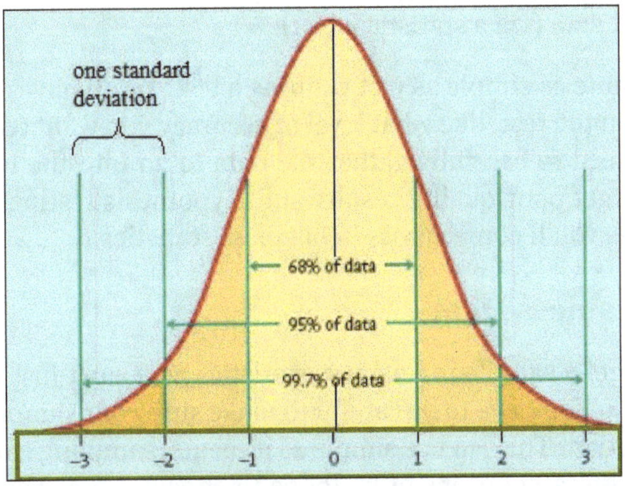

Standard normal distribution.

Of all these, the normal distribution (the "bell curve") is probably the most recognizable and is widely used in business. Businesses use the model for lots of reasons, including compensation and performance reviews. While the curve is relatively easy to understand and use, caution should be used when applying it to people: New research is showing that the distribution isn't actually a good predictor of people's performance. "As a result," states Forbes, "HR departments and business leaders inadvertently create agonizing problems with employee performance and happiness".

Inferential Statistics

While descriptive statistics "describes" data, inferential statistics make inferences. In other words, you take data and make some sort of conclusion. The two main areas are parameter estimation and hypothesis testing.

Parameter Estimation

With parameter estimation, you take a sample of data (say, 100 out of 1,000) and find a statistic. Let's say you find the average salary of 100 workers is $20 per hour. You take that statistic ($20 per hour) and infer that the population of 1,000 probably has a salary of about $20. It might not be exactly $20, but it's probably going to be close, around $19 to $21. The range is called a confidence interval. This interval is part of the results you get from a hypothesis test.

Hypothesis Test

If you're in business, you might come up with some good ideas about how to improve sales or other metrics. For example, you might think that your average customer is older than average, or that placing certain products at the front of the store can raise the bottom line. These guesses are just shots in the dark without some hard data to back them up. A hypothesis test gives you a solid way to back your guess (called a hypothesis) with some hard data (via a statistical test).

Of course, it isn't quite as simple as just running a test. You'll have a lot of decisions to make before you run the test, like what level of accuracy you want (called a significance level). You'll also need to carefully gather the data to go into the test: If poor quality data goes in, you'll get poor quality results out. Hypothesis testing is one of the most complex procedures you'll come across in business statistics.

Sampling of Business Data

When you want to get a sample in business statistics, you can't just pick a few random items from the stack. You have to be careful to make sure your sample is representative of the entire population. This isn't as simple as it seems; sampling can be a complicated process. There are literally dozens of methods to choose from. These include picking numbers from a hat (called simple random sampling) and using your contacts to make other contacts (called snowball sampling).

Simple Linear Regression and Correlation

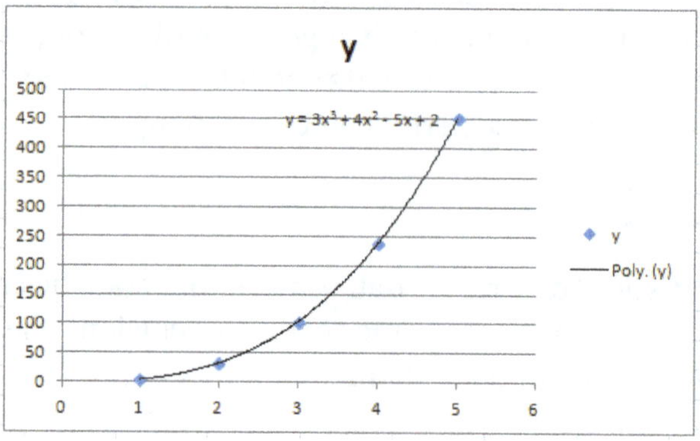

Regression fits data to a line.

Simple linear regression and correlation are used in business statistics to predict trends. For example, you might have a list of sales data from a group of stores. You can use that data to make predictions about where the sales are headed; regression can create a model for the future sales. Correlation takes a set of variables and tells you how well they are related.

Time Series Analysis

A time series graph (also called a timeplot), plots values against time. They are exactly the same as a basic x-y graph with one restriction: the x-axis can only display time.

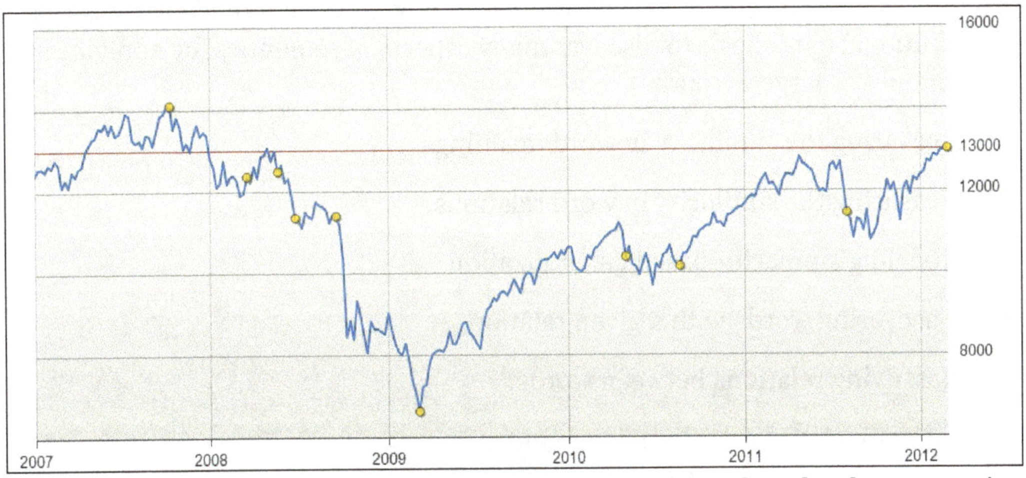

A Dow Jones Timeplot from the Wall Street Journal shows how the stock market changes over time.

Use of Index Numbers in Economic Data

Index numbers tell you how something performs over time. The index always starts at a given year at 100%. Over time, the percentage increases over 100%. For example, an index of 190% shows an increase of 90% from the base year. A decrease shows a decrease compared to the index year. Let's say the mean wage in 1980 was $20,000 and the index for 1990 shows 110%. This means wages increased by 10%, or $2,000. A decrease to 90% in 1990 would mean that wages dropped, on average, by $2,000.

STATISTICAL SEMANTICS

In linguistics, statistical semantics applies the methods of statistics to the problem of determining the meaning of words or phrases, ideally through unsupervised learning, to a degree of precision at least sufficient for the purpose of information retrieval.

The term *statistical semantics* was first used by Warren Weaver in his well-known paper on machine translation. He argued that word sense disambiguation for machine translation should be based on the co-occurrence frequency of the context words near a given target word. The underlying assumption that "a word is characterized by the company it keeps" was advocated by J.R. Firth. This assumption is known in linguistics as the distributional hypothesis. Emile Delavenay defined *statistical semantics* as the "statistical study of meanings of words and their frequency and order of recurrence". "Furnas et al. 1983" is frequently cited as a foundational contribution to statistical semantics. An early success in the field was latent semantic analysis.

Applications

Research in statistical semantics has resulted in a wide variety of algorithms that use the distributional hypothesis to discover many aspects of semantics, by applying statistical techniques to large corpora:

- Measuring the similarity in word meanings.
- Measuring the similarity in word relations.
- Modeling similarity-based generalization.
- Discovering words with a given relation.
- Classifying relations between words.
- Extracting keywords from documents.
- Measuring the cohesiveness of text.
- Discovering the different senses of words.
- Distinguishing the different senses of words.
- Subcognitive aspects of words.
- Distinguishing praise from criticism.

FORENSIC STATISTICS

Forensic statistics is the application of probability models and statistical techniques to scientific evidence, such as DNA evidence, and the law. In contrast to "everyday" statistics, to not engender bias or unduly draw conclusions, forensic statisticians report likelihoods as likelihood ratios (LR). This ratio of probabilities is then used by juries or judges to draw inferences or conclusions and decide legal matters. Jurors and judges rely on the strength of a DNA match, given by statistics, to make conclusions and determine guilt or innocence in legal matters.

In forensic science, the DNA evidence received for DNA profiling often contains a mixture of more than one person's DNA. DNA profiles are generated using a set procedure, however, the interpretation of a DNA profile becomes more complicated when the sample contains a mixture of DNA. Regardless of the number of contributors to the forensic sample, statistics and probabilities must be used to provide weight to the evidence and to describe what the results of the DNA evidence mean. In a single-source DNA profile, the statistic used is termed a random match probability (RMP). RMPs can also be used in certain situations to describe the results of the interpretation of a DNA mixture. Other statistical tools to describe DNA mixture profiles include likelihood ratios (LR) and combined probability of inclusion (CPI), also known as random man not excluded (RMNE).

Computer programs have been implemented with forensic DNA statistics for assessing the biological relationships between two or more people. Forensic science uses several approaches for DNA statistics with computer programs such as; match probability, exclusion probability, likelihood ratios, Bayesian approaches, and paternity and kinship testing.

Although the precise origin of this term remains unclear, it is apparent that the term was used in the 1980s and 1990s. Among the first forensic statistics conferences were two held in 1991 and 1993.

Random Match Probability

Random match probabilities (RMP) are used to estimate and express the rarity of a DNA profile. RMP can be defined as the probability that someone else in the population, chosen at random, would have the same genotype as the genotype of the contributor of the forensic evidence. RMP is calculated using the genotype frequencies at all the loci, or how common or rare the alleles of a genotype are. The genotype frequencies are multiplied across all loci, using the product rule, to calculate the RMP. This statistic gives weight to the evidence either for or against a particular suspect being a contributor to the DNA mixture sample.

RMP can only be used as a statistic to describe the DNA profile if it is from a single source or if the analyst is able to differentiate between the peaks on the electropherogram from the major and minor contributors of a mixture. Since the interpretation of DNA mixtures with more than two contributors is very difficult for analysts to do without computer software, RMP becomes difficult to calculate with a mixture of more than two people. If the major and minor contributor peaks can not be differentiated, there are other statistical methods that may be used.

If the DNA mixture contains a ratio of 4:1 of major to minor contributors, a modified random match probability (mRMP) may be able to be used as a statistical tool. For calculation of mRMP, the analyst must first deduce a major and minor contributor and their genotypes based on the peak heights given in the electropherogram. Computer software is often used in labs conducting DNA analysis in order to more accurately calculate the mRMP, since calculations for each of the most probable genotypes at each locus become tedious and inefficient for the analyst to do by hand.

Likelihood Ratio

Sometimes it can be very difficult to determine the number of contributors in a DNA mixture. If the peaks are easily distinguished and the number of contributors is able to be determined, a likelihood ratio (LR) is used. LRs consider probabilities of events happening and rely on alternative pairs of hypotheses against which the evidence is assessed. These alternative pairs of hypotheses in forensic cases are the prosecutor's hypothesis and the defense hypothesis. In forensic biology cases, the hypotheses often state that the DNA came from a particular person or the DNA came from an unknown person. For example, the prosecution may hypothesize the DNA sample contains DNA from the victim and the suspect, while the defense may hypothesize that the sample contains DNA from the victim and an unknown person. The probabilities of the hypotheses are expressed as a ratio, with the prosecutor's hypothesis being in the numerator. The ratio then expresses the likelihood of both of the events in relation to each other. For the hypotheses where the mixture contains the suspect, the probability is 1, because one can distinguish the peaks and easily tell if the suspect can be excluded as a contributor at each locus based on his/her genotype. The probability of 1 assumes the suspect can not be excluded as a contributor. To determine the probabilities of the unknowns, all genotype possibilities must be determined for that locus.

Once the calculation of the likelihood ratio is made, the number calculated is turned into a statement to provide meaning to the statistic. For the previous example, if the LR calculated is x, then the LR means that the probability of the evidence is x times more likely if the sample contains the victim and the suspect than if it contains the victim and an unknown person. Likelihood ratio can also be defined as 1/RMP.

Combined Probability of Inclusion

Combined probability of inclusion (CPI) is a common statistic used when the analyst can not differentiate between the peaks from a major and minor contributor to a sample and the number of contributors can not be determined. CPI is also commonly known as random man not excluded (RMNE). This statistical calculation is done by adding all the frequencies of observed alleles and then squaring the value, which yields the value for probability of inclusion (PI). These values are then multiplied across all loci, resulting in the value for CPI. The value is squared so that all the possible combinations of genotypes are included in the calculation.

Once the calculation is done, a statement is made about the meaning of this calculation and what it means. For example, if the CPI calculated is 0.5, this means that the probability of someone chosen at random in the population not being excluded as a contributor to the DNA mixture is 0.5.

CPI relates to the evidence (the DNA mixture) and it is not dependent on the profile of any suspect. Therefore, CPI is a statistical tool that can be used to provide weight or strength to evidence when no other information about the crime is known. This is advantageous in situations where the genotypes in the DNA mixture can not be distinguished from one another. However, this statistic is not very discriminating and is not as powerful of a tool as likelihood ratios and random match probabilities can be when some information about the DNA mixture, such as the number of contributors or the genotypes of each contributor, can be distinguished. Another limitation to CPI is that it is not usable as a tool for the interpretation of a DNA mixture.

Blood Stains

Blood stains are an important part of forensic statistics, as the analysis of blood drop collisions may help to picture the event that had previously gone on. Commonly blood stains are an elliptical shape, because of this blood stains are usually easy to determine the blood droplets angle through the formula "$\alpha = arcsin\ d/a$". In this formula 'a' and 'd' are simply estimations of the axis of the ellipse. From these calculations, a visualization of the event causing the stains is able to be drawn, and alongside further information such as the velocity of the entity that caused such stains.

SURVEY METHODOLOGY

A field of applied statistics of human research surveys, survey methodology studies the sampling of individual units from a population and associated techniques of survey data collection, such as questionnaire construction and methods for improving the number and accuracy of responses to surveys. Survey methodology includes instruments or procedures that ask one or more questions that may or may not be answered.

Researchers carry out statistical surveys with a view towards making statistical inferences about the population being studied, and such inferences depend strongly on the survey questions used. Polls about public opinion, public-health surveys, market-research surveys, government surveys and censuses are all examples of quantitative research that use survey methodology to answer questions about a population. Although censuses do not include a "sample", they do include other aspects of survey methodology, like questionnaires, interviewers, and non-response follow-up techniques. Surveys provide important information for all kinds of public-information and research fields, e.g., marketing research, psychology, health-care provision and sociology.

A single survey is made of at least a sample (or full population in the case of a census), a method of data collection (e.g., a questionnaire) and individual questions or items that become data that can be analyzed statistically. A single survey may focus on different types of topics such as preferences (e.g., for a presidential candidate), opinions (e.g., should abortion be legal?), behavior (smoking and alcohol use), or factual information (e.g., income), depending on its purpose. Since survey research is almost always based on a sample of the population, the success of the research is dependent on the representativeness of the sample with respect to a target population of interest to the researcher. That target population can range from the general population of a given country to specific groups of people within that country, to a membership list of a professional organization, or list of students enrolled in a school system. The persons replying to a survey are called respondents, and depending on the questions asked their answers may represent themselves as individuals, their households, employers, or other organization they represent.

Survey methodology as a scientific field seeks to identify principles about the sample design, data collection instruments, statistical adjustment of data, and data processing, and final data analysis that can create systematic and random survey errors. Survey errors are sometimes analyzed in connection with survey cost. Cost constraints are sometimes framed as improving quality within cost constraints, or alternatively, reducing costs for a fixed level of quality. Survey methodology is both a scientific field and a profession, meaning that some professionals in the field focus on survey errors empirically and others design surveys to reduce them. For survey designers, the task involves making a large set of decisions about thousands of individual features of a survey in order to improve it.

The most important methodological challenges of a survey methodologist include making decisions on how to:

- Identify and select potential sample members.
- Contact sampled individuals and collect data from those who are hard to reach (or reluctant to respond).
- Evaluate and test questions.
- Select the mode for posing questions and collecting responses.
- Train and supervise interviewers (if they are involved).
- Check data files for accuracy and internal consistency.
- Adjust survey estimates to correct for identified errors.

Selecting Samples

The sample is chosen from the sampling frame, which consists of a list of all members of the population of interest. The goal of a survey is not to describe the sample, but the larger population. This generalizing ability is dependent on the representativeness of the sample, as stated above. Each member of the population is termed an element. There are frequent difficulties one encounters while choosing a representative sample. One common error that results is selection bias. Selection bias results when the procedures used to select a sample result in over representation or under representation of some significant aspect of the population. For instance, if the population of interest consists of 75% females, and 25% males, and the sample consists of 40% females and 60% males, females are under represented while males are overrepresented. In order to minimize selection biases, stratified random sampling is often used. This is when the population is divided into sub-populations called strata, and random samples are drawn from each of the strata, or elements are drawn for the sample on a proportional basis.

Modes of Data Collection

There are several ways of administering a survey. The choice between administration modes is influenced by several factors, including:

- Costs,
- Coverage of the target population,
- Flexibility of asking questions,
- Respondents' willingness to participate,
- Response accuracy.

Different methods create mode effects that change how respondents answer, and different methods have different advantages. The most common modes of administration can be summarized as:

- Telephone,
- Mail (post),
- Online surveys,
- Personal in-home surveys,
- Personal mall or street intercept survey,
- Hybrids of the above.

There are several different designs, or overall structures, that can be used in survey research. The three general types are cross-sectional, successive independent samples, and longitudinal studies.

Cross-sectional Studies

In cross-sectional studies, a sample (or samples) is drawn from the relevant population and studied once. A cross-sectional study describes characteristics of that population at one time, but cannot give any insight as to the causes of population characteristics because it is a predictive, correlational design.

Successive Independent Samples Studies

A successive independent samples design draws multiple random samples from a population at one or more times. This design can study changes within a population, but not changes within individuals because the same individuals are not surveyed more than once. Such studies cannot, therefore, identify the causes of change over time necessarily. For successive independent samples designs to be effective, the samples must be drawn from the same population, and must be equally representative of it. If the samples are not comparable, the changes between samples may be due to demographic characteristics rather than time. In addition, the questions must be asked in the same way so that responses can be compared directly.

Longitudinal Studies

Longitudinal studies take measure of the same random sample at multiple time points. Unlike with a successive independent samples design, this design measures the differences in individual participants' responses over time. This means that a researcher can potentially assess the reasons for response changes by assessing the differences in respondents' experiences. Longitudinal studies are the easiest way to assess the effect of a naturally occurring event, such as divorce that cannot be tested experimentally. However, longitudinal studies are both expensive and difficult to do. It's harder to find a sample that will commit to a months- or years-long study than a 15-minute interview, and participants frequently leave the study before the final assessment. This attrition of participants is not random, so samples can become less representative with successive assessments. To account for this, a researcher can compare the respondents who left the survey to those that did not, to see if they are statistically different populations. Respondents may also try to be self-consistent in spite of changes to survey answers.

Questionnaires

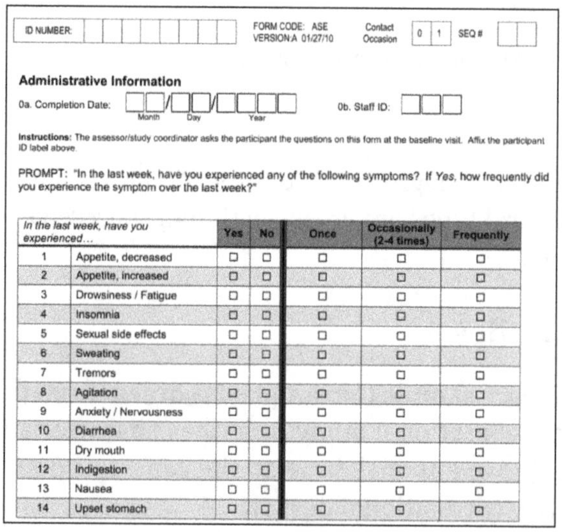
ID NUMBER: FORM CODE: ASE VERSION:A 01/27/10 Contact Occasion 0 1 SEQ #

Administrative Information

0a. Completion Date: Month / Day / Year 0b. Staff ID:

Instructions: The assessor/study coordinator asks the participant the questions on this form at the baseline visit. Affix the participant ID label above.

PROMPT: "In the last week, have you experienced any of the following symptoms? If *Yes*, how frequently did you experience the symptom over the last week?"

In the last week, have you experienced…		Yes	No	Once	Occasionally (2-4 times)	Frequently
1	Appetite, decreased	☐	☐	☐	☐	☐
2	Appetite, increased	☐	☐	☐	☐	☐
3	Drowsiness / Fatigue	☐	☐	☐	☐	☐
4	Insomnia	☐	☐	☐	☐	☐
5	Sexual side effects	☐	☐	☐	☐	☐
6	Sweating	☐	☐	☐	☐	☐
7	Tremors	☐	☐	☐	☐	☐
8	Agitation	☐	☐	☐	☐	☐
9	Anxiety / Nervousness	☐	☐	☐	☐	☐
10	Diarrhea	☐	☐	☐	☐	☐
11	Dry mouth	☐	☐	☐	☐	☐
12	Indigestion	☐	☐	☐	☐	☐
13	Nausea	☐	☐	☐	☐	☐
14	Upset stomach	☐	☐	☐	☐	☐

Questionnaires are the most commonly used tool in survey research. However, the results of a particular survey are worthless if the questionnaire is written inadequately. Questionnaires should produce valid and reliable demographic variable measures and should yield valid and reliable individual disparities that self-report scales generate.

Questionnaires as Tools

A variable category that is often measured in survey research are demographic variables, which are used to depict the characteristics of the people surveyed in the sample. Demographic variables include such measures as ethnicity, socioeconomic status, race, and age. Surveys often assess the preferences and attitudes of individuals, and many employ self-report scales to measure people's opinions and judgements about different items presented on a scale. Self-report scales are also used to examine the disparities among people on scale items. These self-report scales, which are usually presented in questionnaire form, are one of the most used instruments in psychology, and thus it is important that the measures be constructed carefully, while also being reliable and valid.

Reliability and Validity of Self-report Measures

Reliable measures of self-report are defined by their consistency. Thus, a reliable self-report measure produces consistent results every time it is executed. A test's reliability can be measured a few ways. First, one can calculate a test-retest reliability. A test-retest reliability entails conducting the same questionnaire to a large sample at two different times. For the questionnaire to be considered reliable, people in the sample do not have to score identically on each test, but rather their position in the score distribution should be similar for both the test and the retest. Self-report measures will generally be more reliable when they have many items measuring a construct. Furthermore, measurements will be more reliable when the factor being measured has greater variability among the individuals in the sample that are being tested. Finally, there will be greater reliability when instructions for the completion of the questionnaire are clear and when there are limited distractions in the testing environment. Contrastingly, a questionnaire is valid if what it measures is what it had originally planned to measure. Construct validity of a measure is the degree to which it measures the theoretical construct that it was originally supposed to measure.

Composing a Questionnaire

Six steps can be employed to construct a questionnaire that will produce reliable and valid results. First, one must decide what kind of information should be collected. Second, one must decide how to conduct the questionnaire. Thirdly, one must construct a first draft of the questionnaire. Fourth, the questionnaire should be revised. Next, the questionnaire should be pretested. Finally, the questionnaire should be edited and the procedures for its use should be specified.

Guidelines for the Effective Wording of Questions

The way that a question is phrased can have a large impact on how a research participant will answer the question. Thus, survey researchers must be conscious of their wording when writing survey questions. It is important for researchers to keep in mind that different individuals, cultures, and subcultures can interpret certain words and phrases differently from one another. There are two different types of questions that survey researchers use when writing a questionnaire: free response questions and closed questions. Free response questions are open-ended, whereas closed questions are usually multiple choice. Free response questions are beneficial because they allow the responder greater flexibility, but they are also very difficult to record and score, requiring extensive coding. Contrastingly, closed questions can be scored and coded more easily, but they diminish expressivity and spontaneity of the responder. In general, the vocabulary of the questions should be very simple and direct, and most should be less than twenty words. Each question should be edited for "readability" and should avoid leading or loaded questions. Finally, if multiple items are being used to measure one construct, the wording of some of the items should be worded in the opposite direction to evade response bias.

A respondent's answer to an open-ended question can be coded into a response scale afterwards, or analysed using more qualitative methods.

Order of Questions

Survey researchers should carefully construct the order of questions in a questionnaire. For questionnaires that are self-administered, the most interesting questions should be at the beginning of the questionnaire to catch the respondent's attention, while demographic questions should be near the end. Contrastingly, if a survey is being administered over the telephone or in person, demographic questions should be administered at the beginning of the interview to boost the respondent's confidence. Another reason to be mindful of question order may cause a survey response effect in which one question may affect how people respond to subsequent questions as a result of priming.

The following ways have been recommended for reducing nonresponse in telephone and face-to-face surveys:

- Advance letter: A short letter is sent in advance to inform the sampled respondents about the upcoming survey. The style of the letter should be personalized but not overdone. First, it announces that a phone call will be made, or an interviewer wants to make an appointment to do the survey face-to-face. Second, the research topic will be described. Last, it allows both an expression of the surveyor's appreciation of cooperation and an opening to ask questions on the survey.

- Training: The interviewers are thoroughly trained in how to ask respondents questions, how to work with computers and making schedules for callbacks to respondents who were not reached.
- Short introduction: The interviewer should always start with a short introduction about him or herself. She/he should give her name, the institute she is working for, the length of the interview and goal of the interview. Also it can be useful to make clear that you are not selling anything: this has been shown to lead to a slightly higher responding rate.
- Respondent-friendly survey questionnaire: The questions asked must be clear, non-offensive and easy to respond to for the subjects under study.

Interviewer Effects

Survey methodologists have devoted much effort to determining the extent to which interviewee responses are affected by physical characteristics of the interviewer. Main interviewer traits that have been demonstrated to influence survey responses are race, gender, and relative body weight (BMI). These interviewer effects are particularly operant when questions are related to the interviewer trait. Hence, race of interviewer has been shown to affect responses to measures regarding racial attitudes, interviewer sex responses to questions involving gender issues, and interviewer BMI answers to eating and dieting-related questions. While interviewer effects have been investigated mainly for face-to-face surveys, they have also been shown to exist for interview modes with no visual contact, such as telephone surveys and in video-enhanced web surveys. The explanation typically provided for interviewer effects is social desirability bias: Survey participants may attempt to project a positive self-image in an effort to conform to the norms they attribute to the interviewer asking questions. Interviewer effects are one example survey response effects.

ROLE OF STATISTICS IN RESEARCH

The role of statistics in research is to function as a tool in designing research, analysing its data and drawing conclusions therefrom. Most research studies result in a large volume of raw data which must be suitably reduced so that the same can be read easily and can be used for further analysis. Clearly the science of statistics cannot be ignored by any research worker, even though he may not have occasion to use statistical methods in all their details and ramifications. Classification and tabulation, as stated earlier, achieve this objective to some extent, but we have to go a step further and develop certain indices or measures to summarise the collected/classified data. Only after this we can adopt the process of generalisation from small groups (i.e., samples) to population. If fact, there are two major areas of statistics viz., descriptive statistics and inferential statistics. Descriptive statistics concern the development of certain indices from the raw data, whereas inferential statistics concern with the process of generalisation. Inferential statistics are also known as sampling statistics and are mainly concerned with two major type of problems:

- The estimation of population parameters.
- The testing of statistical hypotheses.

The important statistical measures that are used to summarise the survey/research data are:

- Measures of central tendency or statistical averages.
- Measures of dispersion.
- Measures of asymmetry (skewness).
- Measures of relationship.
- Other measures.

Amongst the measures of central tendency, the three most important ones are the arithmetic average or mean, median and mode. Geometric mean and harmonic mean are also sometimes used.

From among the measures of dispersion, variance, and its square root—the standard deviation are the most often used measures. Other measures such as mean deviation, range, etc. are also used. For comparison purpose, we use mostly the coefficient of standard deviation or the coefficient of variation.

In respect of the measures of skewness and kurtosis, we mostly use the first measure of skewness based on mean and mode or on mean and median. Other measures of skewness, based on quartiles or on the methods of moments, are also used sometimes. Kurtosis is also used to measure the peakedness of the curve of the frequency distribution.

Amongst the measures of relationship, Karl Pearson's coefficient of correlation is the frequently used measure in case of statistics of variables, whereas Yule's coefficient of association is used in case of statistics of attributes. Multiple correlation coefficient, partial correlation coefficient, regression analysis, etc., are other important measures often used by a researcher. Index numbers, analysis of time series, coefficient of contingency, etc., are other measures that may as well be used by a researcher, depending upon the nature of the problem under study.

References

- Business-statistics: statisticshowto.datasciencecentral.com, Retrieved 10 July, 2019
- Fung, Wing Kam (2006). "On Statistical Analysis Of Forensic DNA: Theory, Methods And Computer Programs". Forensic Science International. 162 (1–3): 17–23. doi:10.1016/j.forsciint.2006.06.025. PMID 16870375
- Statistics-in-research, research-methodology-tutorial-11500: wisdomjobs.com, Retrieved 14 August, 2019
- Groves, R.M.; Fowler, F. J.; Couper, M.P.; Lepkowski, J.M.; Singer, E.; Tourangeau, R. (2009). Survey Methodology. New Jersey: John Wiley & Sons. ISBN 978-1-118-21134-2

Permissions

All chapters in this book are published with permission under the Creative Commons Attribution Share Alike License or equivalent. Every chapter published in this book has been scrutinized by our experts. Their significance has been extensively debated. The topics covered herein carry significant information for a comprehensive understanding. They may even be implemented as practical applications or may be referred to as a beginning point for further studies.

We would like to thank the editorial team for lending their expertise to make the book truly unique. They have played a crucial role in the development of this book. Without their invaluable contributions this book wouldn't have been possible. They have made vital efforts to compile up to date information on the varied aspects of this subject to make this book a valuable addition to the collection of many professionals and students.

This book was conceptualized with the vision of imparting up-to-date and integrated information in this field. To ensure the same, a matchless editorial board was set up. Every individual on the board went through rigorous rounds of assessment to prove their worth. After which they invested a large part of their time researching and compiling the most relevant data for our readers.

The editorial board has been involved in producing this book since its inception. They have spent rigorous hours researching and exploring the diverse topics which have resulted in the successful publishing of this book. They have passed on their knowledge of decades through this book. To expedite this challenging task, the publisher supported the team at every step. A small team of assistant editors was also appointed to further simplify the editing procedure and attain best results for the readers.

Apart from the editorial board, the designing team has also invested a significant amount of their time in understanding the subject and creating the most relevant covers. They scrutinized every image to scout for the most suitable representation of the subject and create an appropriate cover for the book.

The publishing team has been an ardent support to the editorial, designing and production team. Their endless efforts to recruit the best for this project, has resulted in the accomplishment of this book. They are a veteran in the field of academics and their pool of knowledge is as vast as their experience in printing. Their expertise and guidance has proved useful at every step. Their uncompromising quality standards have made this book an exceptional effort. Their encouragement from time to time has been an inspiration for everyone.

The publisher and the editorial board hope that this book will prove to be a valuable piece of knowledge for students, practitioners and scholars across the globe.

Index

www.ingramcontent.com/pod-product-compliance
Lightning Source LLC
LaVergne TN
LVHW080640210926
848485LV00003B/299